Food Processing Technology

Food Processing Technology

Atul Agnihotri

Food Processing Technology

© Reserved

Edition : 2026

ISBN 978-81-8342-437-0

Published by:

CRESCENT PUBLISHING CORPORATION
4806/24, Mathur Lane,
Ansari Road, Darya Ganj,
New Delhi - 110002
Mob.: 9711991838, 9999021668
Fax : 011-23257835
e-mail : crescentbook@gmail.com
Website : www.crescentpublishingcorp.weebly.com

Printed at :
Roshan Offset Printers
Delhi

Contents

Preface

The role of food processing becomes critical since food production is targeted to double in the next 10 years. With low farmer price realizations and significant wastage in the food supply chain even with the current level of production, only processing of food can secure farmer incomes against a slump in prices as well as reduce wastages. Further, a vibrant food processing industry can be a catalyst for crop diversification.

The impact of increased economic growth in agribusiness through food processing can play a significant role in reducing rural poverty and increasing rural incomes. Food processing leads to significant employment generation, not only directly but also indirectly across the supply chain in production and procurement of raw materials and distribution of food products to consumers.

The study of process engineering is an attempt to combine all forms of physical processing into a small number of basic operations, which are called unit operations. Food processes may seem bewildering in their diversity, but careful analysis will show that these complicated and differing processes can be broken down into a small number of unit operations. For example, consider heating, of which innumerable instances occur in every food industry. There are many reasons for heating and cooling-for example, the baking of bread, the freezing of meat, the tempering of oils.

In food process engineering, heat transfer is very often in the unsteady state, in which temperatures are changing and materials are warming or cooling. Unfortunately, study of heat flow under these conditions is complicated. In fact, it is the subject for study in a substantial branch of applied mathematics, involving finding solutions for the Fourier equation written in terms of partial differentials in three dimensions. There are some cases that can be simplified and handled by elementary methods, and also charts have been prepared which can be used to obtain numerical solutions under some conditions of practical importance.

This is the fundamental reason for writing this book at the present time, to indicate the wealth of knowledge becoming available on reactions in food processing, and the use of reaction technology to apply this knowledge in food processing.

– *Author*

Chapter 1

Food Processing Industries

INTRODUCTION

One of the most important challenges facing the country is providing remunerative prices to farmers for their produce without incurring the additional burden of subsidies. This challenge could be addressed if the processing level and value addition of the agricultural raw produce can be enhanced to meet the growing demand for processed foods. Food processing has an important role to play in linking Indian agriculture to consumers in the domestic and international markets. The role of food processing becomes critical since food production is targeted to double in the next 10 years. With low farmer price realizations and significant wastage in the food supply chain even with the current level of production, only processing of food can secure farmer incomes against a slump in prices as well as reduce wastages. Further, a vibrant food processing industry can be a catalyst for crop diversification.

The impact of increased economic growth in agribusiness through food processing can play a significant role in reducing rural poverty and increasing rural incomes. Food processing leads to significant employment generation- not only directly but also indirectly across the supply chain in production and procurement of raw materials and distribution of food products to consumers.

CONDITIONS AND TRENDS OF CONSUMPTION

In feeding stuffs very great quantities of fats disappear but, with a few unimportant exceptions, they always do so as a constituent of some element of the feed and not as such. Thus the amounts of fat that disappear with grain and legumes fed to animals are truly enormous. In food, as in feeding stuffs, very large quantities of fats are consumed incidentally in the ingestion of meats, fish, poultry, milk, and other animal products and in the form of fats and oils naturally contained in cereals, legumes, fruits, vegetables, and other plant products. However, in the human dietary fats and oils are also consumed practically as such, for example as butter, salad oils, and cooking fats.

QUANTITATIVE DATA UNSATISFACTORY

Not only because fats and oils disappear both in the arts and in the dietary in a vast variety of ways, but for other important reasons, the quantitative study of the consumption of fats and oils encounters great difficulties. In the first place, the very concept of consumption is elusive. Presumably one should exclude, in spite of its potential significance, the fat content of large numbers of meat animals that perish from natural causes and from which no attempt is made to recover the fats.

Commercial consumption is a significant concept, but it excludes a large amount of fats in products that do not enter into trade, and a large amount that are incidental components of meats and other products that do enter into trade, for example, tankage and garbage. Wastes in the process of recovery are considerable; hence the fat content of the product treated may be considerably greater than the fats recovered as such plus those incidentally retained. The amounts ingested by animals and by human beings would be worth knowing, but they would exclude large quantities that are wasted in various ways on the farms and in households and public eating places. Furthermore, a distinction may be drawn between intermediate and final consumption. Large amounts of fats are contained in cereals, nuts, milk, oil cake, and other products fed to animals; these are in part used up by the animals, and in part are converted or stored up and appear in dairy products or slaughtered animals. Hence a total of the fats consumed by animals and those derived from animals would contain, in effect, considerable duplication. The difficulties are enhanced by the fact that our statistical information is defective especially in respect to the data on animal fats.

For the United States, the number of animals killed in inspected slaughterhouses is known; the total slaughter at wholesale is reported by the census of manufactures; but only a guess can be made at the number of animals in other uninspected, principally rural, slaughter. Moreover, there is no clear-cut idea of the fat content per average animal slaughtered; the long-time trend seems to be downward, because younger and lighter animals are coming to slaughter in response to the increasing demand for lamb instead of for mutton, for bacon-type hogs instead of for the lard type, and for baby beef instead of for three- to four-year-old steers. To take even the best ratios derived from packing-house practice and apply them to the total wholesale slaughter of cattle, calves, sheep, and hogs as reported in the biennial censuses, and also to the farm and retail slaughter for which the data are officially admitted to be crude estimates, would be little better than intelligent guessing. At the best, there are gaps in the information so large as to make statistical conclusions on several important components, and on fats and oils as a whole, exceedingly unreliable.

It would be of interest to know, for each of the several sources of fats:

- The amount wasted for lack of any attempt at recovery;
- The amount ingested;
- The amount wasted in households and public eating places;
- The amount fed to animals; and
- The amount used in industrial production other than food.

We should like these data for fats produced, fats imported, and fats exported. We should like to distinguish between gross and net consumption. Unfortunately, for many of the sources of fats the statistical information is exceedingly defective and it is difficult to reach reliable bases upon which to make the estimates necessary for arriving at supplementary data.

We have made attempts to reach a rough approximation to the amounts of fats and oils available for human consumption or industrial use in the United States in recent years, but at several points gaps are so wide and the procedure is so open to objection that we do not feel justified in presenting here even a preliminary approximation, lest in spite of reservations and qualifications it should be taken for more than it would be worth.

In the discussion that follows, therefore, we have incorporated quantitative data only to a very limited extent. The two principal items omitted, fat of dressed meats and milk fats (including butter), are so difficult to estimate that we have not included them; yet their sum probably exceeds the sum of the fats and oils here itemized. They go largely into edible uses, except as milk is fed to animals, though in considerable measure meat fats are wasted in the household or subsequently recovered from household wastes for industrial use. The principal other items omitted are fats incidentally consumed in fish, poultry, grains, vegetables, and fruits. Judging from estimates of Raymond Pearl for the period, the sum of these items is probably less than a billion pounds a year, exclusive of the fat in grain fed to cattle.

FATS AND OILS IN DIET

The function of fats and oils in the diet is mainly to furnish energy to operate the animal machine. In the body, fats are burned as truly as though they were burned in a candle or under a steam boiler, and the end-products of the combustion are the same—carbon dioxide and water. Moreover, the amount of energy they furnish is the same, namely from 9 to 9.4 calories to the gram, whether they are burned within or outside an animal body. (A large calorie, which is the one here used, is the quantity of heat necessary to warm 1 kilogram of water from 0 deg Centigrade to 1 deg Centigrade.

A small calorie is the quantity of heat necessary to warm 1 gram of water from 0 deg Centigrade to 1 deg Centigrade.) They yield more energy than the other important classes of foodstuffs, such as carbohydrates and protein (albumin) Conditions and Trends of Production—Vegetable oils major products), which furnish from 3.8 to 4.2 calories to the gram. Neither fats nor carbohydrates ingested are wholly burned at once unless they are needed for the operation of the animal machine.

If not so needed, carbohydrates are for the most part completely changed in character by conversion into the fat characteristic of the species. This is stored in the body as a reserve against the possibility of a future period of food shortage. Indeed, most of the fat of domesticated animals is produced from starch. Thus the lard of the hogs of the corn belt is mainly derived from the starch of corn (maize).

It has the characteristics of fat normal to the animal, is normally stiff, and hence is especially prized. Unlike ingested carbohydrate, the fat of the food, if it is not at once burned, is changed comparatively little. For the most part it is deposited in but slightly modified form with other fat, made by the animal from carbohydrate or protein, in the storehouses for fat—the adipose tissue under the skin, in and about the viscera, and elsewhere. Therefore, if the food fat differs in its properties from the fat natural to the animal and if there is a great deal of it in the feed, the storage fat formed by the deposition of food fat gives an abnormal character to the fat which is obtained from the animal after slaughter.

The dietary fats may be divided into four groups:

1. Those consumed as such on the table;
2. Those employed in the preparation of foods;
3. Those consumed incidentally in the ingestion of meats and other animal products; and
4. Those consumed incidentally in cereals, legumes, fruits, and vegetables.

The fats consumed as such on the table in the United States are principally butter, butter substitutes (chiefly coconut and oleo oil), and salad oils (principally cottonseed, corn, and olive oils). The fats and oils used in the preparation of food are butter, lard, lard compounds (for the most part cottonseed oil), cottonseed oil as such, corn oil, peanut oil, butter substitutes, and, in certain types of confectionery, coconut oil and cacao butter. Fats and oils consumed incidentally in the ingestion of meats and dairy products are found principally in milk, cheese, poultry, eggs, beef, mutton and lamb, pork, and fish.

The fat of meats is of course to a material extent not ingested but wasted except in so far as it is recovered in garbage and similar greases. The fats of the cereals, legumes, fruits, and vegetables as a class are important in the aggregate nutritionally, but have little direct commercial importance, since few foodstuffs are purchased on the basis of fat content. Practically all foodstuffs except sugar, water, and certain condiments contain some fat. Fresh fruits and vegetables contain merely traces; nuts, on the other hand, are rich in fats, as are also chocolate products. Oat meal and corn meal that is made without removing the germ are relatively rich in fat; wheat flour is poorer.

A study of the trends of consumption encounters the difficulty that statistics of fat in the diet over a series of decades are not available. Certain inferences seem nevertheless warranted. Trends of consumption may be grouped in two classes according as they are due primarily to changes in habits or primarily to substitutions made by manufacturers of which the consumer may or may not be aware. The two classes of course overlap, for in certain cases both factors play a part. Thus a manufacturer may create a new product which leads to a new food habit or, vice versa, consumer demand may lead to substitution or creation of a new product by the manufacturer.

INFLUENCE OF CHANGES IN FOOD HABITS

Change of food habits may be of two sorts. It may be purely quantitative or it may be qualitative. By a quantitative change is meant increase or decrease in consumption of a given fat rather than the substitution in the diet of one fat for another. By qualitative change is meant primarily a substitution of one fat for another. However, since for physiological reasons the intake of food is practically constant for any individual under any given set of circumstances, a decrease in fat ingestion usually involves either a substitution for it of some other kind of food, or of some other kind of fat. Conversely, an increase usually involves a decrease in consumption of some other kind of food or of fat. Since all fats and oils have very nearly the same food value, one can be substituted for another without change in the volume of food ingested. This statement applies to energy values. Certain fats contain small quantities of chemical substances of unknown chemical nature but of great importance to health. These are known as vitamins or food accessories.

Different fats contain different amounts of them depending upon their origin, method of preparation, and other factors. From the point of view of their vitamin content all fats are not of equal nutritive value; but these are considerations. However, if another food be substituted, the volume of the diet is thereby necessarily increased since no other food has, weight for weight, so great an energy value. It follows that if the replacement of fat in the diet by other foods goes too far the diet must become very bulky if it is still to furnish the same energy value, and this fact sets mechanical limits to the substitution of other foodstuffs for fat in the diet.

Since the caloric requirements of an individual, as stated above, are constant under given conditions, it follows that under these circumstances a reduction in consumption of a fat involves the substitution for it of either an equal amount of some other fat or a corresponding amount of some other foodstuffs. But conditions do not remain constant for any person. He grows old and requires less food. He may become stouter or leaner, then consuming either more or less food. He may change his means of livelihood from one requiring hard manual labour involving high food requirements to a sedentary occupation requiring a low food intake. He may migrate from a very cold climate where food requirements are relatively high to a hot one where they are relatively low. In any nation all these changes are taking place in some individuals in one direction, in others in the opposite direction. For short periods, the result is a reasonably constant consumption.

Over a long period of time this is no longer true. The age distribution of populations changes. If the birth-rate falls and the average span of life lengthens, the proportion of old people who consume little food increases, the proportion of young people who consume much decreases. If it becomes unfashionable to be stout, less food is consumed by the nation as a whole. If the proportion of manual labourers becomes less and that of machine tenders and sedentary workers greater, the per capita consumption of food tends to fall. Since fat is the most concentrated form of food energy, its intake tends to be reduced rather more than that of other forms of food.

In the United States exactly these changes have been taking place for decades; and there is evidence of a corresponding decline in the nutritional use of fats and oils, as part of a general reduction in the per capita food requirements. This reduction is the result of substitution of machine labour for man labour, the decline of outdoor employment in severe winter weather, the improved heating of buildings, the trend to lower average body weights of the people generally, and changes in the age distribution of the population. The population is acquiring more sedentary characteristics, needs less food, and the average per capita requirements of foodstuffs in terms of calories are automatically reduced. Most of these factors are the result of the economic evolution of the country and in essence reflect improvement in the standard of living. The rise of the standard of living is responsible for certain other changes in food habits. It results in a diversification of the diet with increase in the use of dairy products, fruit, vegetables, and sugar, and decline in the ingestion of cereals and fats. This general statement does not hold for the fat of milk, the use of which is on the increase. Fat consumed incidentally to the ingestion of meats and dairy products has always been heavy. The consumption of milk and poultry fat is increasing, that of beef, sheep, and hog fats is declining. Plain cooking is being replaced by more fancy cooking, which includes more discriminating culinary uses of fats and oils. American practice has departed from the British custom of boiling vegetables without fat. Much of the crackers and bakers' bread consumed now contains shortening.

These facts might seem to suggest increased use of fats and oils. A survey of the entire field, however, suggests that the net result is a lower per capita ingestion of fats in general. The heavy fat rations of hard workers are a thing of the past—fat-backs and sow-belly are no longer staples. We use more butter, but less hog fat. The current public taste for younger and lighter-weight animals represents a substantial reduction in fat. It seems fair to infer that, milk fat and poultry fat aside, the decline in the ingestion of fats applies to animal fats. Over a generation, indeed, it is possible that there may have been an absolute increase in the per capita consumption of vegetable oils with an absolute decline in the per capita consumption of animal fats outside of milk fat.

INFLUENCE OF CHANGES IN FOOD MANUFACTURE

The changes in trends of food-fat consumption are not all by any means due solely to change in habits. Some of those that are quantitatively extremely important are due to substitutions on the initiative of manufacturers of food-fat preparations—for example, the substitution of lard compounds for lard—or to the increasing availability of new types of fat—for example, coconut oil. Some of the possibilities of substitution in food uses have already been touched upon incidentally. However, since dietary consumption is the major and the higher or premium use of fats, the practice and possibility of substitution require elaboration because they are basic to any consideration of the commodity economics of this important group of raw materials. The substitution of one fat for another, indeed the mere possibility of such substitution, naturally has the greatest influence upon prices.

Given sufficient price inducement, one fat may in some cases displace another with the widest repercussions upon the producer, the farmer, the trader, and the manufacturer. It is one of the purposes of the *Fat and Oil Studies of the Food Research Institute* to present studies from time to time upon far-reaching movements of this general character. That considerations of price, costs of conversion and refining, lack of technological skill, and legislation in the interests of public health and sanitation limit the volume of inedible fats turned to edible uses, has already been pointed out.

But there are other factors as well that limit not merely the diversion of inedible fats to edible uses but also the substitution of one edible fat for another. The most prominent of these are greater or lesser adaptability of different fats to use in the preparation of different foodstuffs, and psychological and sentimental factors that in the food industries play a greater part than in the arts.

IMPACT OF ORGANISED RETAILING ON EMPLOYMENT

The restriction of FDI in the retail sector is based on the premise that entry of large international players will displace existing kirana stores and impact employment. However, there is empirical evidence to suggest that development of large retail formats will translate into greater employment as witnessed in other countries such as China, Japan, Singapore and Malaysia.

The sector is labour intensive and contributes significantly to employment. The current share of organised retailing and share of retailing in total employment is tabulated below. Higher share of organised retailing has led to higher share of retailing in total employment. In India, the share of employment of organized food retailers is insignificant, on account of the limited size of this segment at present. Traditional kirana outlets employ approximately 20 million people. However, there is significant shadow employment at these outlets, reflecting that the real employment potential of these outlets is far lower.

INDIA'S POSITION IN WORLD TRADE.

Global agricultural exports (including food products) increased from USD 326 billion in 1990 to USD 520 billion in 2003. Strong expansion in food and agri exports in the mid 1990s, was followed by a decline from 1997 to 2000.This was mainly due to a slump in prices for various agricultural commodities, on the back of demand-supply mismatches. Since 2001,food and agri trade has been growing. The main items in food and agri trade include fruits and vegetables (-USD 75 billion),cereals and cereal-based preparations (-USD 58 billion), marine products (-USD 57 billion), meat and meat-based preparations (-USD 46 billion), milk and milk products (-USD 33 bn) and beverages (-USD 40 billion).Latin America has by far the largest agricultural trade surplus, followed by Oceania. Brazil and Argentina contribute significantly to the Latin American trade surplus. Asia has the largest trade deficit of all regions, primarily on account of Japan. Western and Eastern Europe as well as Africa have a trade deficit too.

About half of global trade flows are regionally contained, partly as a result of regional agreements and partly due to the perishable nature of agri-commodities. There has been a sharp increase in regional trade agreements (RTAs) over the past decade. A total of 259 RTAs have been notified to the WTO by the end of 2002, although only 176 RTAs are operational. An additional 70 RTAs are estimated to be operational, but not yet notified by WTO and about 70 are under negotiation. A well known example of regional trade is in the European Union, where about 75% of agricultural trade is within the region. The intra-regional trade figures for other regions are 33% (NAFTA),63% (Asia) and 15% (Latin America and Africa).

The group of G-20 countries, a grouping of developing countries led by Brazil, China, India and South Africa have sought further policy reforms and market liberalization on the part of the developed world. At the Geneva meeting of the WTO, the members have agreed on a further reduction of overall trade distorting domestic support, the elimination of all forms of export subsidies and a reduction of import tariffs. The EU already anticipated this by drastically reforming the Common Agricultural Policy (CAP). Besides agricultural policies and trade arrangements, demand-supply trends and the macroeconomic environment also significantly influence trade. Another important factor is the change in currency rates.

EXPORTS

India's exports of agricultural and food products are about INR 360 billion (approx. USD 8 billion) which constitutes about 1.6% of total global trade (USD 520 billion). Exports of agriculture & food products have grown at 15% annually, vis a vis total exports which have grown at 19%, in the last decade. The share of agricultural exports as a percentage of India's total exports has decreased from 19 % to 13 % in this period. Excluding non-food agricultural products (such as paper, cotton and jute), tobacco and seeds; marine exports constitute the single largest product category in agricultural and food exports from India followed by, rice oilmeals, wheat, cashew, tea and coffee. Most of the agricultural products are primary processed. Many products have shown negative or single digit growth. The top 15 export categories in food and agricultural (except for non-food items and tobacco) from India are listed below. With the exception of buffalo meat, rice and cashew, India's share in global agricultural and food trade is insignificant. Marine products which have the highest share in Indian exports, have a 2% share of global trade in this category. Among the key food exports, only non-basmati rice, wheat and buffalo meat has shown double-digit growth. Marine products, basmati rice and processed fruit juices/vegetables have shown single digit growth. Products like cashew, tea, spices, coffee have shown negative growth.

IMPORTS

Against exports of approximately USD 8 bn India's import of agricultural products are about USD 5 bn which translates into trade surplus of approximately USD 3 bn (Year 2003-04). India has abolished restrictions on imports of many agricultural and food products post liberalisation. However, tariff quotas are maintained on some edible oils, maize and milk powder. There are restrictions on imports of certain fats, oils of animal origin, and beef, based on GATT Article XX that permits countries to restrict imports on religious grounds. Some products, such as wheat, rye, oats, maize, rice, canary seed and other cereals, continue to be traded by the Food

Corporation of India. Despite liberalisation of imports, India's imports are growing at about 1 % per annum for the last five years.

While exports cover a wide range of products, imports are skewed towards pulses and oilseeds. Pulses and oilseeds are the key food items in India's imports (73% share) of food items displaying year on year growth. Increasing production of edible oils in India can reduce India's dependence on imports to a large extent. Another category which has witnessed growth in imports is Scotch whisky. The volume growth is about 13% and value growth is about 6% in the last decade. However, the growth is driven by imports of bulk Scotch, which is bottled in India and/or blended with Indian liquor by domestic manufacturers. Bottled Scotch imports have not witnessed any growth due to the current level of import duties.

ISSUES HINDERING EXPORTS

India continues to be either absent or at best a marginal player in most of the leading markets for its exports. Indian players have not succeeded in establishing direct linkages with buyers/consumers in importing countries, as a result of which a large proportion of exports are being further processed and re-exported by other countries.

PRODUCT

Marine products - Decline in raw material availability

- Fragmented base of suppliers
- Lack of availability of technology *e.g.* for tuna fishing
- High duties on imports of additives/flavourings for making value added products
- Anti-dumping issues (such as excess use of antibiotics) in key markets

Rice

- Outdated milling technology (using rubber roll sheller) resulting in high share of broken rice
- Poor quality of seeds/non-availability of certified authentic seeds leading to lack of consistency in grain quality
- Low cost competitiveness due to state taxes and MSP regime (for non-basmati]

Cashew

- Lack of raw material availability - high dependence on imports of raw nuts, Interstate barriers
- on movement of raw nuts
- Purchase tax for exports in some states (*e.g.*4 % in Tamil Nadu]
- Lack of mechanization of processing leading to higher production costs
- Lack of established quality parameters for raw cashew

Key Issues Hindering Exports

Tea- Export market viewed as a 'residual market' to sell surplus production.

- High cost of production
- Production focused on CTC rather than orthodox, whereas demand for orthodox tea is growing faster than for CTC

Coffee- Low realizations due to exports restricted to green coffee and no exports in roasted/instant/branded coffee.

Export cess of INR 500/MT leading to low profitability for exporters

BUFFALO MEAT

Mango

- Absence of policy on permitting rearing for slaughter which translates into lack of traceability a key requirement in several importing countries
- Inappropriate facilities at existing municipal slaughter houses.
- License for setting up private slaughter house is difficult to obtain.
- Prevalence of various diseases particularly Foot and Mouth Disease (FMD)
- Lack of exportable varieties (high fibre content; inappropriate appearance and texture and large size of stone)
- Lack of post-harvest treatment facilities such as for vapor treatment
- Lack of packhouses from farm to port

Grapes

High cost of setting up vineyards

- High cost for obtaining certification for exports. For *e.g.*, the cost of EurepGap certificate is
- INR. 75000/farmer (including the cost of construction of separate storage space for fertilizers and pesticides etc)

Dairy Products

Lack of quality monitoring mechanism in the supply chain as required by many importing countries

- Opportunistic production of SMP with low exportable surplus
- Few local manufacturing facilities of value added products
- Lack of appropriate packaging technology

Guar Gum

Variation in production from year to year (high share of rain-fed areas).

Increasing domestic demand from non-food applications such as textiles, paper, pharmaceuticals, oil well drilling.

UNECONOMIC SCALE OF OPERATIONS

Lack of consistency in supply and quality

Lack of cost competitiveness due to statutory charges, intermediation and wastages/losses

Inadequate and inappropriate storage and distribution infrastructure

NON TARIFF BARRIERS SHORT PRODUCT LIFE CYCLE LACK OF BRAND IMAGE

Most exporters from India lack scale- for example the largest fresh produce exporter records annual sales of about INR500mn.The low volume translates into lack of economies in operations and makes exports uncompetitive. Hence, exporters are not able to establish themselves as long-term players in the export market, and rely heavily on opportunistic businesses. These factors cumulatively translate into low investments in upgrading skill sets, product innovation, quality improvement and brand building.

Low economies of scale high cost and low volume.

Low investment capacity in product innovation and Brand/ Market development.

Low competitiveness in global markets unable to establish as a long-term player.

Low efficiencies technological and managerial.

Countries across the world have successfully addressed these issues to become globally competitive.

RICE, CASHEW AND COFFEE EXPORTS FROM VIETNAM

Rice Increased competitiveness through higher productivity: High competitiveness in rice has been achieved on the back of consistent increase in area and productivity for paddy. Cropping patterns have been adjusted to increase land area for planting winter-spring paddy and summer-autumn paddy (from 2.1 and 1.2 million ha in 1990 to 2.89 and 2.35 million ha in 2000 respectively) and reducing acreage under low - yielding winter paddy (from 2.74 to 2.4 million ha).

Intensive farming and more advanced technologies have led to a consistent increase in paddy yields from 3.69 tons/ha in 1995 to 4.6 tons/ha in 2003; higher than other Asian countries including India (2 tons/ha),Thailand (2.2 tons/ha), Myanmar (3.2 tons/ha) and the Philippines (2.89 tons/ha).

These measures have facilitated the increase in output from 19.2 million tons in 1990 to 31.3 million tons in 1999, and further to 34.4 million tons in 2003.

Cashew - Mechanization of processing technology: Vietnam is the largest cashew producer in South East Asia and the third largest cashew exporter in the world after India and Brazil. Over 90% of production is exported. The number of cashew exporters have increased from 16 in 1997, to over 50 in 2003.

The cashew processing industry in Vietnam has made significant contributions to enhance exports. The developments in processing technology have enabled Vietnam to export cashew in processed form. Vietnam has developed "Cover split technology-designed by Vietnamese technicians. This technology is cheap and is able to generate a higher ratio of whole seed. Due to easy availability of efficient technology, the number of processing companies increased from 6 in 1986 to 30 in 1994 (with total capacity of 75000 tons/ year) and to 62 in 1999 (with total capacity of 250000 tons/year) to about 120 in 2003.

Coffee - Government Support: Vietnam is ranked as the second largest coffee exporter in the world since 2000, next to Brazil. It exports coffee to about 62 countries. The two largest markets include the European Union (47 % share) and the USA (15 % share).The Government has played an important role in increasing coffee exports. Some of the initiatives include:

Inviting foreign investment in production and trading of coffee - many of the world's largest trading houses in the world such as ED & F Man, Newman Groupe and O Lam are present in Vietnam. Promoted application of international standards to domestic coffee.

In order to cope with the market changes, the Government and industry bodies carried out studies to determine the appropriate proportion of Robusta and Arabica to be cultivated in line with global demand.

Accordingly, Robusta plantations were replaced by Arabica.The aim is to achieve a proportion of Arabica and Robusta as 75:25,to align it with global demand.

EMPLOYMENT GENERATION POTENTIAL OF FOOD PROCESSING INDUSTRIES

Food Processing has significant potential for employment generation not only directly but across the supply chain in production of raw materials, storage of produce and finished products and distribution of food products. For *e.g.* a grant of INR 66.7 million (total investment of approximately INR 250 to 300 million) to 35 units in UP in 2003-04 has resulted in direct employment of 2,500 and indirect employment of 20,000, with a significant rural component.

Employment intensity is significantly higher in the Small Scale Industries (SSI) sector as compared to the organized sector for the same level of investment.

The incremental employment in the organized sector in FPI sector by 2015 on the basis of the stated vision is estimated at 8.2 Million. The sectoral break-down is as follows.

The employment intensity in the organised sector is 1.8 direct and 6.4 indirect per million of investment. The ratio of indirect to direct employment is therefore 3.5. Rabo India has estimated investment required in the organised sector of FPI as INR 997 bn in next ten years. Hence the employment generation potential in the orgnaised sector is 8.2 million including 1.8 million direct and 6.4 million indirect for an investment of INR 997 bn.

The unorganized sector in food processing requires an investment of about INR 100 bn in next ten years (estimated on the basis of output ratio as 2:1 and capital intensity ratio as 5:1 of organized and unorganized sector. The employment intensity is estimated to be approximately 10 direct employment per INR million of investment in the unorganised sector (Source: Dr.JS Bedi Analysis).This will lead to direct employment creation of 1 million in unorganized sector.

The indirect employment generation in the unorganised sector will be about 1 million (assuming ratio of direct to indirect employment as 1:1).The above analysis assumes no replacement of existing employment.

SSI IN FOOD PROCESSING

The SSI Sector accounts for 95% of industrial units in the country,40% of value added in the manufacturing sector, 34% of national exports and 7% of Gross Domestic Product (GDP).The SSI sector is the largest employment generator next only to Agriculture. It has been estimated that an investment of INR 1 million in fixed assets in the small-scale sector generates employment for forty persons and produces more than four million rupees worth of goods or services. The Food sector is a leading employer within SSI, providing employment to 480,000 persons (13% of SSI).

The SSI sector is less capital intensive with a high potential to generate employment. However, the efficiency of SSI units is impacted by the following:

- Lack of capital/credit
- Inadequate Training technical/managerial
- Tools and technology (traditional and less efficient)
- Limited market knowledge (demand, food standards)
- SSI's role in semi-processing: The organized large-scale

sector is focused on processed foods, where SSI cannot compete due to lack of marketing and distribution strengths. However, SSIs can play an important role in procuring from farmers and primary processing of produce to increase shelf life and make it available to processor/marketers who have access to the final consumer.

- Need for product innovation and branding: There is a strong need to provide necessary training and R&D support to SSIs to promote product innovation. Also, SSIs have limitations in terms of investments on brand development. There is a need to promote public private participation in supporting collective investment by SSIs in branding.

FINANCING NEEDS OF THE AGRICULTURAL AND FOOD SUPPLY CHAIN

The Indian agribusiness supply chain is highly fragmented, with independent players engaged in various value-addition activities such as input supplies, cultivation, processing, distribution, storage and retailing. The scale of operations of each entity is limited not only due to the structure of the supply chain, but also on account of Government policies as well as due to demand-related factors.

In contrast, internationally, a large number of food companies are integrated across the chain. *e.g.* players such as ADM and Cargill have business interests in agricultural inputs, procurement, transportation and storage of produce, processing and value addition.

The structural complexion of the Indian supply chain thus translates into limited scale of financing as well as higher risk, given the lack of control of each of the players, on the supply chain.

ISSUES WITH FARMER FINANCING

It is important to understand the issues faced in farmer level financing, both from the borrower's as well as the lender's perspective, and develop solutions which address these issues comprehensively.

Farmer's Perspective - Limited Access to Credit

Non-availability of timely credit from organized sources, mainly on account of the processes and procedures of banks/financial institutions.

This compels farmers to rely on intermediaries.

- Inadequacy in credit availability-The 'scale of finance' method, stipulated by RBI, caps total credit availability per farmer.
- Inability to offer tangible security/collateral cover to access credit from banks Subsistence farmers, who constitute a large proportion of India's farming community, are unable to benefit from the organized credit system since they cannot offer adequate tangible security. Further, the security offered (land documents) are notional and cannot be enforced by banks.
- The above factors, thus translate into continued reliance on intermediaries for financing. While intermediary financing is timelyas they do not require the level of documentation as banks, the cost of funds is exorbitant, nearly 4 times that of interest rates charged by banks.
- A more efficient agri-supply chain requires that farmers' financing needs are addressed by banks and linkages of farmers with middlemen are scrapped.

Banker's Perspective

- Lack of credit discipline among borrowers Lack of credit discipline, to some extent is infused by the Government's decision to waive outstanding loans of farmers. Policy-related uncertainties translate into a huge risk for banks. Instead of loan waivers, the Government should seek to impose a short-term moratorium on repayments, to enable farmers to overcome the adverse impact of a poor monsoon/ crop failure.
- Lack of tangible security from farmers- as mentioned above, farmers are unable to provide tangible security to financiers.
- Lack of linkages of farmers with processors- In absence of forward linkages with markets, banks run a huge performance risk, as there is greater potential for farmer default.
- Inadequate insurance cover the existing crop insurance schemes are inadequate, and this further increases the risk profile of farmers

- Norms for priority sector lending- As per the current RBI policy on priority sector lending, kharif credit is not accounted for in priority sector lending. This results in a lop-sided focus on extending financing to farmers, favouring crops cultivated in the rabi season.

Policy-related Issues

- Cooperatives Act: There are about 95,000 Primary Agriculture Credit Co-operatives (PACS) and about 100,000 marketing co-operatives. As per the Cooperative Act, cooperatives are allowed to borrow from Cooperative Banks, DCCBs and RRBs. This provision needs to be extended to scheduled banks, in order to enhance flow of credit to farmers.
- State Warehousing Corporations Act: Private sector scheduled banks are constrained from lending to State Warehousing Corporations. Given the sizeable investment required to create grain storage infrastructure, it would be in the interest of the CWC/SWCs that a wider participation of lenders is permitted.
- Restrictions on land holdings: The State Land Holding Laws prohibit consolidation of land holdings.

Financing to farmers with larger land holdings, and thus larger scale of operations, could result in higher credit flows from banks. Further, the Government could also consider consolidation of wastelands and leasing of these to corporates. This would not only enable increase in financing to agriculture, but also aid crop diversification.

CHALLENGE FACING THE FOOD INDUSTRY

The challenge facing the food industry, and this is linked to shifts in the food system more generally, is how to make healthy foods more marketable, and marketable foods appear more healthy. Both questions are capable of eliciting a response from food manufacturers, but to understand the background to this response we need to examine the patterns of resistance which have grown up to recent changes in the food industry, patterns which echo some of the forms of resistance to environmental. The beginnings of what Belasco calls a 'countercuisine' are linked to a number of quite important social changes which, as we have already noted, shift the individual's attention from the sphere of production relations to those of consumption. Three aspects require our attention: the way in which we arrive at our impressions of what foods to avoid and what foods to consume (the quality of food as a commodity in the market); the way in which food 'signifies' and expresses other aspects of our lives (food as self-enhancement); and finally, the way in which our consumption of food is linked to broader questions of ownership and organization in the economy as a whole (food as part of a broader political economy).

Almost unnoticed, issues surrounding each of these aspects of the consumption of food have assumed enormous importance in recent years, but close examination reveals that what at first appear to be almost random evidence of social resistance to the modern food system have their roots in fundamental social changes. These social changes need to be outlined, and their implications examined. It is our contention that too much attention has been devoted to the 'signifying' aspects of food and diet, the semiology of food consumption, by sociologists and anthropologists, to the detriment of a wider understanding of the social transformations implied by the economic and technological changes in the food system.

As we have seen, women's roles have been transformed in a number of ways, some of which are clearly irreversible. Domestic work, including housework, has been opened up to important market forces, and the manufacture and purchase of 'white goods' for the home is now an important activity. At the same time the idea that housework is no longer onerous has served the interests of manufacturers and others. It is far from clear that technology in the home is labour-saving; this assumption, needs to be examined critically. Evidence exists that a human price is exacted for the benefits of speed and convenience which 'white goods' bring. Housework standards change, and women expect (and are expected) to raise the standards of housework in turn. At the same time, it is likely that women's release from the worst forms of labour drudgery is associated with more time being spent by them on other activities in the home, particularly attention to children. This is not to say that many household tasks are not easier; they are. It is merely to point out that the definition of housework, and with it women's responsibilities in the home, is constantly shifting. It is not immutable. It is also clear that since the 1970s important changes have occurred in popular understanding of the relationship between food and health. Preventative medicine may only be in its infancy, and receive little official encouragement, but for some groups of people in the industrialized world, healthy eating is now considered essential. Healthy eating in the past depended critically on local custom and diet and, most importantly, income. Today this is still true, but local variations have less importance, and additional factors have made their appearance, which have engaged the attention of a battery of professionals, such as health educators, doctors, consumer groups and alternative therapists. The perceived need to diet to reduce obesity has given rise to a huge consumer market in low-calorie foods, including the controversial 'low-cal' liquid diets and innumerable diet 'systems'.

The publication of evidence about heart disease and cancer has led to some marked shifts in the public consumption of certain foods, especially in the United States, where the changes date back to the publication of the McGovern Report in 1977. Food has become one of the obsessions of our time, associated as it is with good health, longevity and the occurrence of stress. Physical fitness has also become an obsession in some quarters, perhaps also a 'fad'. Evidence from the United States suggests that, while in 1960 only a quarter of adults exercised regularly, by 1980 over half did so.

The point is not that everybody is exercising more, but that more people are, and that health is increasingly linked to diet in many people's minds. This is not a trend likely to be discouraged by the food industry, since it also represents a change with commercial possibilities. Indeed, the giant food firms have jumped on this bandwagon, adapting to each new nutritional recommendation—low sodium, low saturated fats, high fibre—to the point of confusing consumers grappling with labels and elusive media-speak, such as 'lite' and 'natural'.

Changes have also occurred in public perceptions of the two processes: appropriation and substitution. Following the scare over DDT between 1969 and 1972 in the United States, and subsequent publicity about the use of other pesticides, attention has been given to the effects of pesticide residues in food, especially by the London Food Commission in the United Kingdom. Few studies have been undertaken, and even fewer given publicity, about the relative nutritional quality of organically produced and chemically produced food. Pesticide use in the United States rose 500 per cent between 1950 and 1986, but in the latter year one-fifth of United States's crops were lost to pests, the *same percentage as in* 1950! The tactic employed by the food industry in meeting the criticism of environmentalists and whole food enthusiasts, has been that the industry can 'compensate' through industrial processes for the shortcomings of nature.

The industry argued that organic foods were more expensive (ignoring the hidden subsidies provided to chemically produced food), that organic foods could not meet market demand, that food additives increased the palatability of food, and that convenience removed the drudgery from food preparation. By the late 1960s in North America, and perhaps a decade later in the United Kingdom, food manufacturers began to 'reposition' themselves, as well as their food products. Now they were on the side of emancipated women and environmentalists; they offered themselves as accomplices in women's drive for more independence and in the vanguard of responsible, sustainable resource management. Market researchers have credited much of the improvement in the food industry to their own responsiveness to consumer pressure. In the words of an industry spokesperson:

For example, ten years ago the only place people could find additive-free *natural* products were in the comparatively cramped and small premises of health food stores. Today such products, not only much improved and—dare we say it?—engineered to satisfy consumer preferences, are mainstream products in prime placing in major supermarket chains the length and breadth of Britain.

This represents one side of the story, but in practice healthier eating has also presented a challenge to the food industry which it has found difficult to grasp. The problem, in a nutshell, is how to leave out food additives and processing, and still add value to the final product.

The post-war period had seen additives 'substitute' for natural ingredients, while much that was natural had been removed from food. Food was given more value added through packaging, processing and ensuring it stayed 'fresh' longer. The challenge represented by making 'healthy' foods from 'health' foods was how to make profits by appearing to do less to the product. The industry responded to the challenge, as we have noted, by both market and product differentiation—identifying, and exploiting market opportunities for 'niche' products, such as low-calorie foods, ethnic health foods and fitness-related foods.

Finally, the food industry has been in the forefront of the promotion of food supplements; people can now supplement their inadequate diet from a range of commercially promoted food 'accessories'. In 1985 over US$3 billion was spent in the United States on vitamin and mineral supplements and over $4.5 billion in supplements to fortified breakfast cereals. This diet supplementation was the inevitable outcome of the way eating had changed in the United States. In 1909, 40 per cent of calories had been provided from fruit, vegetables and grains. By 1976 only 20 per cent of calories came from these sources, the rest from fats and refined sugars. As people ate more calories they ate fewer vitamins and minerals—supplements have helped to fill the vacuum left by changes in the 'affluent diet' of North America.

Another important social change that has accompanied the shifts in diet referred to above is that governments in the industrialized countries have been forced to make some accommodation to the pressures put upon them by food consumers, and interest groups representing consumers. In the United Kingdom this has come about largely as a result of successive food 'scares', such as those over salmonella, listeria and bovine spongiform encephalopathy ('mad cow disease') in the last few years. The extent to which governments will respond to consumer anxieties is still unclear, but the current debate about public nutrition policy and food legislation represents an excellent illustration of the kind of contradiction to emerge from the development of the modern food system.

ROLE OF LEGISLATION AND FOOD CONTROL

The primary purposes of food legislation are to protect the health of the consumer and to protect the consumer from fraud. In the case of fortified foods, there is a need to ensure that the population is not at risk of receiving toxic doses of any micronutrient. Food laws must also ensure that the target population does not receive nutritionally ineffective levels of micronutrients. Procedures for monitoring premises where fortified foods are prepared, packed, stored or held for sale as well as mechanisms for penalising defaulters must be clearly defined within the food regulations.

In Switzerland samples of fortified products, both local and imported, are taken from the market annually for analysis of vitamin content by two specialised institutes. If vitamin content is found to be outside of established acceptable limits, the official government agency can disallow sale of the product.

Standards for fortified foods and labelling requirements must also be contained within the food regulations. Standards play an important role in the facilitation of trade, both nationally and internationally. When large differences exist between national standards, however, they can forn technical barriers to trade. In light of the Agreement on Technical Barriers to Trade, the development of international standards for fortified foods is an important step in the elimination of technical barriers.

It has been found that food law is managed most effectively in two parts: a basic food act and food regulations. The act itself should set out broad principles while the regulations should contain the detailed provisions governing the different categories of products. Within the regulations should be found lists of approved fortificant compounds and food standards stating the allowed levels of nutrients in the fortified foods.

This organization gives some flexibility to food law as it is much more difficult to have laws amended than to revise regulations. Prompt revision of regulations may become necessary because of new scientific knowledge, changes in new processing technology or emergencies requiring quick action to protect the public health.

With respect to regulations dealing with fortified foods, changes might be prompted as a result of safety evaluations on nutrient compounds or new information regarding the roles and optimal levels of specific micronutrients in the maintenance of good health. Changes in food processing and packaging technologies could be shown to result in a significant reduction in processing and storage losses of micronutrients, thus requiring a revision in the allowed levels of addition of nutrients. In the face of demonstrated micronutrient deficiencies, regulations regarding standards for certain foods and levels of fortification may need to be revised.

CONSUMPTION OF FOOD PRODUCTS

The consumption of food in India was estimated at approximately INR 4,000 bn at 1993-94 prices (INR 7,250 bn at 2001-02 prices) in the year 2001 -02, growing at a CAGR of 7.8% (1996-2002). Food consumption has been driven primarily by increase in per capita expenditure on food items. In the period 1996-2002, the per capita expenditure grew by 6 % per annum and population increased at the rate of 1.6% per annum. Urban food consumption has grown at 9.1 % annually and is estimated at INR 1,540 bn (1993-94 prices). Rural food consumption, on the other hand, has grown at 7% annually and is estimated at INR 2,550 bn (1993-94 prices). Uttar Pradesh is the largest consumer market for food products followed by Maharashtra, West Bengal, Bihar and Andhra Pradesh. Among the fastest growing markets are the North East states, Gujarat and Andhra Pradesh.

The share of grain-based products is the highest in the Indian consumer's food basket, followed by milk and milk products, vegetables, edible oil and meat products. The growth rates for fruit, vegetables, meat and dairy products is higher than grains and pulses, indicating a shift in consumption pattern. This also implies the need for diversification in agricultural production to match the change in consumption patterns.

CONSUMPTION OF PROCESSED FOOD PRODUCTS

The market size for processed foods for year 2003-04, at current prices is estimated at INR 4,600 bn, which is about 53 % of total food consumption. This includes processing in unorganized sector in dairy (Halwais) and grains (chakkis).The market size excluding these segments is about INR 3,300 bn. The market size of processed foods at factory cost for 2003-04 is estimated at INR 2,100 bn[3] implying about 60 % mark-up from factory cost to market price.

However, primary processed products constitute as much as 62 % of processed foods consumed, with value-added products being the balance 38%.

Exhibit 1: Estimated Total Food Market in 2003-04 (INR bn) at current prices

Total food Consumption*	8,600
Processed foods**	4,600
Primary processed food (includes packed	2,800
packed milk, unbranded edible oil,	
sugar, pulses, spices and salt)	
Value-added foods (includes processed	1,800
juices, jams, pickles, squashes,	
products - ghee, paneer, cheese, butter,	
branded edible oil; breads, biscuits,	
foods; processed meat, poultry and	
and chocolates; alcoholic beverages - beer,	
and malted beverages)	

'excluding consumption of alcoholic beverages and out-ofhome consumption

"excluding out-of-home consumption but including alcoholic beverages and processing in unorganized sectors in dairy (halwais) and grain milling (chakkis)

Source: Rabobank Analysis

The low share of value added products in food consumption is due to low level of processing across agricultural products.

Exhibit 2: Level of Processing in Perishable Products

Products		**Level of**	
	Organi	Unorgan	Total
Fruits and	1.4%	0.8%	2.2%
Milk & milk	13%	22%	35%
Meat and Poultry			
Buffalo	21%	-	21%
Poultry	6%	-	6%
Marine	8%	-	8%

The key impediments to growth of processed foods include demand-side, supply-side and regulatory issues.

The demand side factors include:

- low per capita income
- socio-cultural factors such as preference for freshly cooked products as compared to packaged products and availability of low-cost domestic help

The supply side factors constraining the demand for processed foods include:

- long and fragmented supply chain with independent players engaged in various value-addition activities such as input supplies, cultivation, processing, distribution, storage and retailing (unlike in other major agriculture/food producing countries, where a large number of food companies are integrated across the chain. *e.g.* ADM, Cargill, Tyson Foods)
- high cost of raw material (due to low productivity and poor agronomic practices) and lack of varieties amenable for processing
- lack of scale
- high cost of packaging
- lack of infrastructure (cost of power, lack of cold chains, storage, handling and transportation)
- high cost and poor quality of distribution.

FOOD ACIDITY AND PROCESSING METHODS

Whether food should be processed in a pressure canner or boiling-water canner to control botulinum bacteria depends on the acidity of the food. Acidity may be natural, as in most fruits, or added, as in pickled food. Low-acid canned foods are not acidic enough to prevent the growth of these bacteria.

Acid foods contain enough acid to block their growth, or destroy them more rapidly when heated. The term "pH" is a measure of acidity; the lower its value, the more acid the food. The acidity level in foods can be increased by adding lemon juice, citric acid, or vinegar.

Low-acid foods have pH values higher than 4.6. They include red meats, seafood, poultry, milk, and all fresh vegetables except for most tomatoes. Most mixtures of low-acid and acid foods also have pH values above 4.6 unless their recipes include enough lemon juice, citric acid, or vinegar to make them acid foods. Acid foods have a pH of 4.6 or lower. They include fruits, pickles, sauerkraut, jams, jellies, marmalades, and fruit butters.

Although tomatoes usually are considered an acid food, some are now known to have pH values slightly above 4.6. Figs also have pH values slightly above 4.6. Therefore, if they are to be canned as acid foods, these products must be acidified to a pH of 4.6 or lower with lemon juice or citric acid.

Properly acidified tomatoes and figs are acid foods and can be safely processed in a boiling-water canner. Botulinum spores are very hard to destroy at boiling-water temperatures; the higher the canner temperature, the more easily they are destroyed. Therefore, all low-acid foods should be sterilised at temperatures of 240° to 250°F, attainable with pressure canners operated at 10 to 15 PSIG.

PSIG means pounds per square inch of pressure as measured by gauge. The more familiar "PSI" designation is used in (the Complete Guide to Home Canning). At temperatures of 240° to 250°F, the time needed to destroy bacteria in low-acid canned food ranges from 20 to 100 minutes.

The exact time depends on the kind of food being canned, the way it is packeu into jars, and the size of jars. The time needed to safely process low-acid foods in a boiling-water canner ranges from 7 to 11 hours; the time needed to process acid foods in boiling water varies from 5 to 85 minutes.

PROCESS ADJUSTMENTS AT HIGH ALTITUDES

Using the process time for canning food at sea level may result in spoilage if you live at altitudes of 1,000 feet or more. Water boils at lower temperatures as altitude increases. Lower boiling temperatures are less effective for killing bacteria. Increasing the process time or canner pressure compensates for lower boiling temperatures. Therefore, when you use the Complete Guide to Home Canning, select the proper processing time or canner pressure for the altitude where you live.

NUTRITION VALUE

Canning is a way of processing food to extend its shelf life. The idea is to make food available and edible long after the processing time. Although canned foods are often assumed to be of low-nutritional value (due to heating processes or the addition of preservatives), some canned foods are nutritionally superior-in some ways-to their natural form. For instance, canned tomatoes have a higher available lycopene content.

POTENTIAL HAZARDS

Migration of Can Components

In canning toxicology, *migration* is the movement of substances from the can itself into the contents. Potential toxic substances that can migrate are lead, causing lead poisoning, or bisphenol A, a potential endocrine disruptor that is commonly use to coat the inner surface of cans.

Botulism

Foodborne botulism results from contaminated foodstuffs in which *C. botulinum* spores have been allowed to germinate and produce botulism toxin, and this typically occurs in canned non-acidic food substances. *C. botulinum* prefers low oxygen environments, and can therefore grow in canned foods. Botulism is a rare but serious paralytic illness, leading to paralysis that typically starts with the muscles of the face and then spreads towards the limbs.

In severe forms, it leads to paralysis of the breathing muscles and causes respiratory failure. In view of this life-threatening complication, all suspected cases of botulism are treated as medical emergencies, and public health officials are usually involved to prevent further cases from the same source.

CHEMICALS IN FOOD PROCESSING

The same over commitment to potentially risky methods is found in other phases of food production. Complex problems of storage, cleansing, handling, refining, cooking, mixing, heating and packaging are often solved by the use of chemical additives. A half century ago the yellowish tint in freshly milled flour was removed slowly by aging; today it is removed rapidly by oxidizing agents. The miller no longer has to store flour and wait for it to mature. Emulsifiers are employed, partly to "improve" the texture of a product, partly to speed up processing. Their "functional advantages" in ice cream, for example, are that the "mix can be whipped faster, the product from the freezer is dryer and the mix has less tendency to 'churn' in the freezer."

Chemical antioxidants are used to help prevent oils and fats from becoming rancid, a problem that becomes especially pronounced when natural antioxidants are removed during large-scale processing. Chemical preservatives are added to food to permit longer storage in wholesale depots and to extend the "shelf life" of a product in retail outlets. In time chemicals are turned from mere adjuncts of food production into "technological necessities." Their use may result in new machines and facilities, the abandonment of old processing methods and a broad reorientation of technology to meet the new chemical requirements.

The fact that a chemical is technologically useful does not mean that it is biologically desirable. The two phrases are occasionally used interchangeably, as though they were synonymous. Some of the chemicals most widely used in the production of food have been found to be harmful to animals in laboratory tests. Nitrogen trichloride, or agene, was employed for decades as a bleaching and maturing agent in flour before it was discovered, in 1946, that the compound produces "running fits," or hysteria, in dogs.

The use of agene in the bread industry has since been discontinued. Although there is no evidence that agenized flour causes nerve disorders in man, no one can say with certainty that cereal products treated with agene are completely harmless to consumers, especially if eaten day after day for twenty years or more. The toxicant in agenized flour "produces nervous disturbances in monkeys, but not epilepsy in man," observed the late Anton Carlson, one of the most distinguished physiologists of his day. "That is not sufficient for me, because there are many other types of nervous disturbances that may be aggravated or introduced by this chemical."

Another group of chemicals, notably certain types of polyoxyethylene derivatives, were widely used in the United States as emulsifiers before they were banned as potential hazards to consumers. Former F.D.A. Commissioner Charles W. Crawford is reported to have described the agents as "good paint removers." According to data furnished to the Delaney Committee, rats that had been fed two commercial preparations of polyoxyethylene-lb/> derived emulsifiers showed blood in their feces and developed a variety of kidney, abdominal, cardiac and liver abnormalities. These preparations were used primarily as bread softeners but also found their way into other foods. The polyoxyethylene derivatives constitute a highly suspect family of emulsifying agents, yet some of them are still added to cake mixes, cake icing, frozen desserts, ice cream, candies, dill pickles, fruit and vegetable coatings, vitamin preparations and many food-packaging materials. Similar examples can be culled from other groups of chemical aids. In general, the toxic effects of many commonly used compounds are difficult to determine, but individually and in groups they have aroused suspicion and concern. "The fortification of oils and fats against oxidation and rancidity is another field in which there are serious questions of possible toxicity," warns F. J. Schlink, technical director of Consumers' Research and editor of *Consumer Bulletin*. "The natural antioxidants which delay rancidity are lost during the factory processing of refined oils and fats; then the attempt is made to restore the antioxidation qualities by addition of fat-stabilizing substances. Most of the materials that have antioxidation properties are known to be toxic to some degree."

The term "technological aid" has been used loosely to include food additives that could be dispensed with entirely if a manufacturer improved his processing methods and the quality of his product. The question of whether a chemical, instead of being legitimately employed to furnish distant consumers with a perishable product, was being used rather as a device for masking inferior ingredients and unsanitary methods arose in sharp form early in the century, when sodium benzoate was employed extensively as a preservative in catsup. Harvey Wiley, the first administrator of the federal food and drug law, regarded sodium benzoate as a hazard to consumers.

Supported by data from feeding experiments performed with human beings, Wiley claimed that the preservative causes injury to the digestive system. He was convinced that catsup need not deteriorate readily if clean facilities and proper ingredients are used in preparing it. Challenged on this point by Charles F. Loudon, a canner in Terre Haute, Indiana, Wiley appointed a bacteriologist, Arvill W. Bitting, to work at the Loudon factory and prepare a formula for a stable catsup preparation without artificial preservatives. Bitting was successful. Moreover, after inspecting numerous canning factories and examining many brands of catsup, Bitting concluded that manufacturers who needed sodium benzoate often used inferior products and were careless about sanitation. "The whole spirit and tradition of the pure-food law is against the use of preservatives and substitutes," declared Mrs. Harvey Wiley more than forty years after the Loudon episode. "The intent of the law is to encourage the manufacture of high-grade ingredients, not to try to hide inferiority and cheapness by the use of chemicals and preservatives." If this is so, the spirit of the law has been ignored for decades.

The most up-to-date surveys of chemical additives in foods show that preservatives are added to cheese, margarine, cereal products, jams, jelly and many other processed foods. Uncooked poultry is dipped in solutions of antibiotics. Aside from the damage to public health that may arise from the use of questionable chemicals as preservatives, the longer "shelf life" acquired by some of these foods may cause a loss of valuable nutrients. Many chemical additives in foods perform no technological or nutritional function whatever. They cannot be regarded as materially useful by any stretch of the imagination. These chemicals enhance the colour of a food or make certain products feel soft and newly prepared; in some instances they have been used to replace costly but valuable nutrients with inferior ones. In the least objectionable cases, an additive will deceive the consumer without impairing his health.

The application of an innocuous vegetable dye to some foods often leads the consumer to believe that he is acquiring a better, more wholesome, or tastier product than is actually the case. In the most objectionable cases, a chemical is toxic to a greater or lesser degree. This type of additive not only serves to change the appearance or conceal the nutritive deficiency of a food; it exposes the consumer to a certain amount of damage.

Nitrates and nitrites, for example, are used to impart a pink colour to certain brands of processed meat. Ostensibly, this is their sole function. The addition of these chemicals to meat, especially frankfurters and hamburger meat, would be objectionable if only because of the deception which makes it difficult for the housewife to distinguish between high-quality and low-quality products. But this is not the only deception perpetrated on the consumer. Used together with salt, nitrite compounds definitely extend the "shelf life" of processed meat products. When nitrites are ingested, they react with hemoglobin in the blood to form methemoglobin and, like carbon monoxide, reduce the hemoglobin's capacity to carry oxygen. An individual who consumes three to four ounces of processed meat containing 200 parts per million of sodium nitrite (a permissible residue) ingests enough of the compound to convert from 1.4 to 5.7 per cent of his hemoglobin to methemoglobin. Ordinarily, this percentage is insignificant. But if the same individual is a heavy smoker and lives in an air-polluted community, the cumulative loss in oxygen-carrying capacity produced by the nitrites in the food and by the carbon monoxide in tobacco smoke and motorcar exhaust can no longer be dismissed as trivial.

Sodium nitrite is highly toxic in relatively small amounts. About four grams of the compound constitutes a lethal dose for adults. Although Lehman regards 200 parts per million of sodium nitrite as a safe residue, he notes that "only a small margin of safety exists between the amount of nitrite that is safe and that which may be dangerous. The margin of safety is even more reduced when the smaller blood volume and the corresponding smaller quantity of hemoglobin in children is taken into account. This has been emphasized in the recent cases of nitrite poisoning in children who consumed weiners and bologna containing nitrite greatly in excess of the 200 ppm permitted by Federal regulations. The application of nitrite to other foods is not to be encouraged."

Emulsifiers have been used in bread not only as softening agents, which can give stale bread the texture of newly baked bread, but as substitutes for nourishing ingredients. "The record of the bread-standards hearings contain evidence of distribution among bakers of advertising material advocating the replacement of fats, oils, eggs and milk by emulsifiers," George L. Prichard, of the Department of Agriculture, told the Delaney Committee. "The use of such products as components of food may work to the disadvantage of our farm economy by displacing farm products normally used. The record indicates that natural food constituents, such as fats and oils, probably will be reduced in many commercial bakery products if bakers are allowed to employ these emulsifiers."

This substitution affects more than 117 the 117 farm 117 economy, however; it also works to the disadvantage of the consumer.

To illustrate this, the Delaney Committee report compared the ingredients in two cake batters prepared by the same company eleven years apart; during the interval emulsifiers were added to the product. The first batter, made in 1939, did not contain synthetic emulsifiers; the second, prepared in 1949, did. "On a percentage basis, the cake batter in 1939 contained 13 per cent eggs and 8.6 per cent shortening.

In 1949, the cake batter contained 6.3 per cent eggs and 4.8 per cent shortening, with somewhat less than 0.3 per cent of synthetic emulsifier." The report noted that a "synthetic yellow dye could be added to provide the colour formerly obtainable through the use of eggs. The utilization of synthetic yellow dye in commercial cake was practiced before the war. There are indications that the use of artificial colouring matter is increased when quantities of whole eggs or egg yolks are reduced in commercial cake formulas." To heighten the insult, among the most commonly used emulsifiers in 1949 were the polyoxyethylene monostearates. The yellow dye referred to in the Delaney Committee report was probably FDC Yellow No. 3(Yellow AB), which, until fairly recently, was added to many yellow cake mixes. The dye often contained impurities of a potent carcinogen and its use in foods was forbidden in 1959. For a number of years, however, both the emulsifier and the dye undoubtedly appeared in many brands of cake and reached large numbers of unsuspecting consumers.

Problems of this kind are not likely to disappear unless there is a basic change in the viewpoint of the F.D.A. "Inherited from the Wiley era is a too common misconception that all 'chemicals' are harmful and the related idea that any amount of a 'poison' is harmful," the F.D.A. observes in a brochure on food additives. "The fact is, of course, that chemical additives, or food *additives* as they are now being called, have brought about great improvements in the American food supply. Additives like potassium iodide in salt and vitamins in enriched food products are making an important contribution to the health of our people and yet it is a fact that both iodine and some of the vitamins would be harmful if consumed in excessive amounts. Many similar examples could be given to refute these common misconceptions."

This argument is grossly misleading. Iodine and vitamins are indispensable to human life, but coal-tar dyes and benzoic acid, for example, are not. If we adhered to a well-balanced diet of properly prepared natural foods, iodine and vitamins would never enter the body in toxic amounts. Coal-tar dyes, on the other hand, are suspected of being harmful to man in nearly any amount if consumed repeatedly and the kindest statement that can be made for the presence of benzoic acid in food is that the compound is "safe under the conditions of its intended use."

The formula "safe under the conditions of its intended use" exposes the consumer to serious risks when it opens the door to the use of food additives whose biochemical activity is not understood. Many unexpected problems may arise when such additives appear in food. A particular additive may seem to be relatively harmless to the organs of the body, but it may be carcinogenic on the cellular level of life. Another additive may produce insignificant or controllable effects when studied in isolation; brought into combination with various chemicals in food, however, it may give rise to toxic compounds. The body, in turn, may make a toxic additive more poisonous in the course of changing it during metabolism. "At one time it was generally believed that whenever a toxic substance was absorbed the body was capable of calling upon special mechanisms for detoxifying the toxicant," Lehman observes. "It was believed also that the metabolic pathway that the toxicant followed always proceeded in the direction of the formation of a less toxic compound. Later work on the metabolism of drugs and toxic substances showed that special mechanisms did not exist, but that the toxicant was subject to the same metabolic influences as those which normally operate in the body. The assumption that the metabolic product was less poisonous than the parent substance from which it was derived is also unwarranted simply because in many instances the toxicity of either the original substance and its conversion product is unknown. In other instances the metabolic product is even more poisonous than its parent. The conversion product, heptachlorepoxide, which is two or three times more poisonous than the parent substance, heptachlor [a widely used insecticide], may be cited as an example of this."

Finally, food additives may cause allergic reactions that are likely to go unnoticed for many years. The reactions need not be severe to be harmful. Otto Saphir and his colleagues at the Michael Reese

Hospital in Chicago have recently suggested that the development of arteriosclerosis may be promoted by the sensitivity of the body to allergenic compounds, notably certain antibiotics. By using sulfa drugs to produce allergies in forty-two rabbits, the researchers were able in eight months to cause degenerative changes in the arteries of thirty-one of the animals.

These changes closely resembled arteriosclerosis in man. According to a press account of Saphir's report, the rabbits' reactions "were not apparent on the surface." It would be very imprudent to assume that such effects are produced only by chemicals that cause severe or noticeable allergies. If the data of Saphir and his colleagues are applicable to man, additives with even minor allergenic properties cannot be dismissed as harmless.

Today more than 3,000 chemicals are used in the production and distribution of commercially prepared food. At least 1,288 are purposely added as preservatives, buffers, emulsifiers, neutralizing agents, sequestrants, stabilizers, anticaking agents, flavouring agents and colouring agents, while from 25 to 30 consist of nutritional supplements, such as potassium iodide and vitamins. The remaining chemicals are "indirect additives"—substances, such as detergents and germicides, that get into food by way of packaging materials and processing equipment. Many chemical additives are natural ingredients, but a large number are not. The artificial substances that appear in food range from simple inorganic chemicals to exotic compounds whose biochemical activity is still largely unknown.

DISTRIBUTION OF FOODS IN INDIA

Food products are sold through over 5 million food and grocery stores in India. The organised food retail market is estimated at INR 20 bn (approximately 0.2 % of food expenditure in India).This is in contrast to developed countries, where food distribution is highly consolidated. The majority of food and food products are retailed through neighborhood kirana stores. The kirana stores focus on dry food products in the absence of infrastructure for cold storage. Bulk of fresh produce is sold by vendors with push carts. Meat, poultry and marine products are primarily sold by small retailers in wet markets. Such produce is associated with low product quality, lack of variety and low hygiene levels.

Internationally, food retailers have played an important role in improving supply chain efficiencies such as developing storage and transportation infrastructure, training supply chain members on food hygiene and standards and providing scientific know how to farmers. The key impediments to growth of organised food retailing in India include lack of infrastructure, technology and capital.

Foreign Direct Investment (FDI) is not allowed in retailing with the exception of cash & carry formats. This restriction is based on the premise that entry of large international players will displace existing employment and reduce bargaining power of farmers. However, there is empirical evidence from other countries (USA, China) which highlights that opening of the retail sector to FDI has enabled employment creation, disintermediation, increased farmer access to information and wider choice and cost savings for the consumer.

EXPORTS OF AGRICULTURAL AND FOOD PRODUCTS

India has 1.5 % (INR 360 bn in 2003-04) share of global agricultural exports (approx USD 522 bn), despite its production leadership in agriculture. India's exports primarily constitutes commodity and primary processed items, where price realisations are low. In addition, many products are showing single digit or negative growth.

The reasons for India's insignificant share in global trade include supply side factors such as lack of consistency in supply and quality, lack of cost competitiveness, and demand side factors such as non-tariff barriers, short product life cycles and perception of Indian food products in overseas markets.

Most exporters from India lack scale- for example the largest fresh produce exporter records annual sales of about INR 500mn.This has resulted in lack of economies in operations and renders them uncompetitive. Hence, exporters are not able to establish themselves as long-term players in the export market, rely on opportunistic businesses and are consequently, unable to develop technical and managerial expertise. These factors cumulatively lead to low investments by exporters in brand building, quality improvement and brand development.

STATUS OF FOOD DISTRIBUTION IN INDIA

Food accounts for the largest share of consumer spending. Food and food products account for about 53% of the value of final private consumption estimated at INR 8600 bn (2003-04 at current prices). Food products are sold through over 5 million food and grocery stores in India. The size of the organised food retail segment in India is estimated at INR 20 bn (approximately 0.2 % of food expenditure).This is in contrast to most of the developed countries, where food distribution is highly consolidated.

In countries such as Australia, France, Germany, Denmark and The Netherlands, the share of top the 10 retailers is more than 80% (USA ~ 35%, China ~ 15%).

Most organised retailers in India are regional and use single formats such as convenience stores, supermarkets, hypermarkets or cash and carry. In contrast, most international retailers have multiple format models. Besides reflecting the stage of evolution of retailing in India, this also highlights the capital constraints of existing players in India, restricting their ability to make large investments.

In India, food companies are much larger in size than the organised food retailers in India. In contrast, the retailers are larger in scale, than food companies, in developed countries. Given the channel power of retailers in these markets, food companies have started establishing partnerships with them to develop and test new products, share consumer information and undertake consumer promotions.

KEY DRIVERS

Increasing need for convenience: The Indian consumer visits about eight to ten outlets to purchase various food products which constitute the daily consumption basket. These outlets include neighbourhood kirana stores, bakeries, fruit and vegetable outlets, dairy booths and chakkies (small flour mills). With changing lifestyles, there is growing paucity of time, and convenience in food shopping is emerging as an important driver of growth of one-stop retail formats.

Availability of quality retail space: Until the late 1990s, the high cost of real estate meant that organised food retail business models were not financially viable in metropolitan areas. In the last few years, various factors have led to increased availability of real estate for organised retail formats. About 300 malls are at various stages of construction, across metros and mini-metros in the country. The average size of a mall is about 100,000 sq.ft. About 25 malls each are planned in Delhi and Mumbai, primarily in suburban and satellite areas. This will translate into additional retail space of 30 to 40 million sq.ft over the next three to five years.

KEY IMPEDIMENTS

The key impediments to organised food retailing include lack of infrastructure, technology and capital.

Lack of infrastructure and technology: Food distribution in India is characterised by a high degree of complexity given dispersed production on the one hand, and variance in demand across locations on the other. There is a compelling requirement for appropriate infrastructure for storage and transportation such as temperature controlled warehouses and vans. However, the limited scale of operations of retailers has restricted their investment capacity in these areas.

The recent entrants who have fuelled growth of organised formats are attempting to address this issue, but given the large requirements of capital, the process of transformation is gradual. Further, entry of organised retailers, who have larger investment capacity, will also address the issue of creation of infrastructure for storage and transportation.

Lack of capital: Organised food retailing is a capital intensive business with a long gestation period (typically 5-7 years). Investment is required in buying/leasing land, furniture and fixtures, IT systems, quality control systems, vendor development and infrastructure for warehousing, storage and transportation.

The development of organised food retailing in India has been constrained due to lack of capital. Foreign Direct Investment (FDI) is not allowed in retailing with the exception of cash & carry formats.

Chapter 2

Basic Principles of Food Processing Engineering

The study of process engineering is an attempt to combine all forms of physical processing into a small number of basic operations, which are called unit operations. Food processes may seem bewildering in their diversity, but careful analysis will show that these complicated and differing processes can be broken down into a small number of unit operations. For example, consider heating of which innumerable instances occur in every food industry. There are many reasons for heating and cooling–for example, the baking of bread, the freezing of meat, the tempering of oils.

But in process engineering, the prime considerations are firstly, the extent of the heating or cooling that is required and secondly, the conditions under which this must be accomplished. Thus, this physical process qualifies to be called a unit operation. It is called 'heat transfer'.

The essential concept is therefore to divide physical food processes into basic unit operations, each of which stands alone and depends on coherent physical principles. For example, heat transfer is a unit operation and the fundamental physical principle underlying it is that heat energy will be transferred spontaneously from hotter to colder bodies.

Because of the dependence of the unit operation on a physical principle, or a small group of associated principles, quantitative relationships in the form of mathematical equations can be built to describe them. The equations can be used to follow what is happening in the process, and to control and modify the process if required.

Important unit operations in the food industry are fluid flow, heat transfer, drying, evaporation, contact equilibrium processes (which include distillation, extraction, gas absorption, crystallization, and membrane processes), mechanical separations (which include filtration, centrifugation, sedimentation and sieving), size reduction and mixing.

These unit operations, and in particular the basic principles on which they depend, are the subject of this book, rather than the equipment used or the materials being processed. Two very important laws which all unit operations obey are the laws of conservation of mass and energy.

HIGH PRESSURE PROCESSING TECHNIQUE

High Pressure Processing (HPP) preserves a food's natural flavour, nutrients, and other sensory properties while extending the shelf life of foods through the inactivation of microorganisms. The uniform application of this non-thermal food processing technology also provides the food industry with new products and new product development opportunities that can fully exploit the functional properties of food ingredients such as hydrocolloids, proteins, etc.

Some of the successful food applications include dressings, sauces, salsas, packaged meats, seafood, cut fruit, purees, juices, chilled ready-to-serve desserts, soups, yogurt, neutraceuticals, pharmaceuticals and other food products. The many advantages of using high pressure processing (HPP) in food production have been known for over a century. However, the technology and equipment required to efficiently and reliability generate the extreme pressures (up to 600 MPa/87,000 psi) used in HPP have only recently become commerciably viable.

Food preservation using high pressure is a promising technique in food industry as it offers numerous opportunities for developing new foods with extended shelf life, high nutritional value and excellent organoleptic characteristics. High pressure is an alternative to thermal processing. The resistance of microorganisms to pressure varies considerably depending on the pressure range applied, temperature and treatment duration, and type of microorganism. Generally, Gram-positive bacteria are more resistant to pressure than Gram-negative bacteria, moulds and yeasts; the most resistant are bacterial spores. The nature of the food is also important, as it may contain substances which protect the microorganism from high pressure.

This chapter presents results of studies involving the effect of high pressure on survival of some pathogenic bacteria—*Listeria monocytogenes, Aeromonas hydrophila* and *Enterococcus hirae*—in artificially contaminated cooked ham, ripening hard cheese and fruit juices. The results indicate that in samples of investigated foods the number of these microorganisms decreased proportionally to the pressure used and the duration of treatment, and the effect of these two factors was statistically significant (level of probability, *P* d" 0.001). *Enterococcus hirae* is much more resistant to high pressure treatment than *L. monocytogenes* and *A. hydrophila*. Mathematical methods were applied, for accurate prediction of the effects of high pressure on microorganisms.

The usefulness of high pressure treatment for inactivation of microorganisms and shelf life extention of meat products was also evaluated. The results obtained show that high pressure treatment extends the shelf life of cooked pork ham and raw smoked pork loin up to 8 weeks, ensuring good micro-biological and sensory quality of the products. Food products are an excellent environment for growth of pathogenic microorganisms, which may cause food-borne diseases.

Quality and shelflife of food products depend greatly on the properties of microorganisms contaminating the food. Despite the introduction of food standards obligatory in EU countries, epidemiologists believe that 75 per cent of food-borne diseases are caused by bacteria.

For this reason, the control of microorganisms is an important aspect of food quality and safety. Many methods of food preservation are used for ensuring microbiological safety, among which high pressure processing (HPP) seems a very promising technique for food industry, as it offers numerous opportunities for developing new shelf life stable foods with extended shelf-life, high nutritional value and excellent organoleptic characteristics—minimally processed but safe for consumers. High pressure is an alternative to thermal processing. The resistance of microorganisms to pressure varies considerably depending on the pressure range applied, temperature and treatment duration, and type of microorganism. As a result of technical progress and government support, the first high pressure processed food products appeared in Japan in the early 1990s.

In Europe, high pressure processing (HPP) of foods was rather at the stage of research or pilot production in the last decade. EU legislation included HPP foods in the "novel food" category. EC Novel Food regulation has introduced a statutory pre-market approval system for novel foods across the whole of the European Union. Recently, rapid progress of HPP toward commercial exploitation has been achieved, but still the process requires close collaboration between researchers, food and equipment manufacturers, as well as proper financial support. In Poland, research on the application of high pressure for food preservation was initiated in 1992 by the High Pressure Research Centre (now Institute of High Pressure Physics) of the Polish Academy of Sciences in collaboration with others Institutes.

The research programmes, sponsored by the State Committee for Scientific Research and by EU funds, were devoted to high pressure processing of fruit products, fruit and vegetable juices, milk and meat products. This chapter presents some of the results concerning the effect of high pressure on the survival of *Listeria monocytogenes, Aeromonas hydrophila* and *Enterococcus hirae* in artificially contaminated food products. This microorganisms could be a source of food-borne diseases.

NEED OF HIGH PRESSURE PROCESSING

High Pressure Processing (HPP) is a method of food processing wherein the food is subjected to elevated pressures (pressures up to 87,000 pounds per square inch or approximately 6000 atmospheres) with or without the addition of heat to achieve microbial inactivation or to alter the food attributes in order to achieve consumer-desired qualities. Pressure is effective in inactivating most of the vegetative bacteria at pressures above 400 MPa. HPP retains food quality, maintains natural freshness, and extends microbiological shelf life. HPP can be used to process both liquid and solid (water-containing) foods.

The process is also called as high hydrostatic pressure processing (HHP) and ultra high-pressure processing (UHP) in the literature.

High pressure processing causes minimal changes in the fresh characteristics of foods by eliminating thermal degradation. Compared to thermal processing, HPP results in foods with fresher taste, better appearance, texture and nutrition. High pressure processing can be conducted at ambient or refrigerated temperatures, thereby eliminating thermally induced cooked off-flavours. The technology is especially beneficial for heat sensitive products.

FUNCTION OF HPP

Most processed foods today are heat processed to kill bacteria. Heat oftentimes diminishes the quality of a product. High pressure processing provides an alternative means of killing bacteria which can cause spoilage or food-borne disease without a loss of sensory quality or nutrients. In a typical HPP process, the product is packaged in a flexible container (pouch or plastic bottle) and is loaded into a high pressure chamber filled with a pressure transmitting (hydraulic) fluid.

The hydraulic fluid in the chamber is pressurized with a pump and this pressure is transmitted through the package into the food itself. Pressure is applied for a specific time, usually 3-5 minutes. The processed product is then removed and stored/distributed in the conventional manner. Because of the uniform manner in which the pressure is transmitted (in all directions simultaneously), food retains its shape, even at extreme pressures. And because no heat is needed, the sensory characteristics of the food are retained without compromising microbial safety.

HPP AT A GLANCE

- High Pressure Processing (HPP) is based on the Le Chatelier principle which states that actions that have a net volume increase will be retarded and actions that have a net volume decrease will be enhanced.
- HPP utilizes isostatic or hydrostatic pressure which is equal from every direction.
- During HPP, foods are subjected to pressures up to 100,000 psi. which destroy pathagenic microor-ganisms by interrupting their cellular functions.
- Within a living bacteria cell, many pressure sensitive

processes such as protein function, enzyme action, and cellular membrane function are impacted by high pressure resulting in the inability of the bacteria to survive. Small macromolecules that are responsible for flavour, odour, and nutrition are typically not changed by pressure.

- HPP is gaining in popularity within the food industry because of its capacity to inactivate pathogenic microorganisms with minimal to no heat treatment, resulting in the almost complete retention of nutritional and sensory characteristics of fresh food without sacrificing shelf life.
- One of the unique advantages of HPP is that pressure transmission is instantaneous and uniform, is not controlled by product size and is effective throughout the entirety of the food item.
- As well, HPP offers several advantages over traditional thermal processing including: reduced process times; minimal heat damage problems; retention of freshness, flavour, texture, and colour; and no vitamin C loss.

Like any other processing method, HPP cannot be universally applied for processing all types of foods. At the moment, HPP is being used in the United States, Europe and Japan on a select variety of high value foods either to extend shelf life or to improve food safety. Some products that are being commercially produced using HPP are cooked ready-to-eat meats, avocado products (guacamole), tomato salsa, applesauce, orange juice and oysters.

HPP cannot yet be used to make shelf-stable versions of low-acid products such as vegetables, milk or soups because of the inability of this process to destroy spores. However, it can be used to extend the refrigerated shelf life of these products and to eliminate the risk of various food-borne pathogens such as *Escherichia coli, Salmonella* and *Listeria.* Acid foods are particularly good candidates for HPP technology.

Another limitation is that the food must contain water and not have internal air pockets. Food materials containing entrapped air such as strawberries or marshmallows would be crushed under high pressure treatment, and dry solids apart from caking do not have sufficient moisture to make HPP effective for microbial destruction. During HPP processing, pressure is uniformly applied around and throughout the food product. For example, a grape placed between fingers can be easily squeezed and broken; this is because the pressure is not applied evenly from all sides simultaneously. On the other hand, if the same grape is squeezed from all sides simultaneously, it will not be crushed. This can be demonstrated by placing a grape inside soda bottle filled with water. By squeezing the bottle, you pressurize the water inside as well as the grape. Yet the grape is not damaged, no matter how hard you squeeze. In the same way, foods processed by high pressure will not be damaged by the applied pressure.

SHELF LIFE OF HPP PROCESSED PRODUCT

In general, HPP can provide shelf lives similar to thermal pasteurization. Pressure pasteurization kills vegetative bacteria and, unless the product is acidic, it requires refrigerated storage. For foods where thermal pasteurization is not an option (due to flavour, texture or colour changes) HPP can extend the shelf-life by 2-3 fold over a non-pasteurized counterpart, and improve food safety. As commercial products are developed, shelf life can be established based on microbiological and sensory testing.

High pressure processed products are commercially available in the United States, European and Japanese retail markets. Examples of high-pressure processed products commercially available in the US include fruit smoothies, guacamole, ready meals with meat and vegetables, oysters, ham, chicken strips, fruit juices, and salsa. Low acid shelf-stable products such as soups are not commercially available yet because of the limitations in killing spores with HPP. This is a topic of current research.

It has been generally known that high pressure has very little effect on low molecular weight compounds such as flavour compounds, vitamins and pigments compared to thermal processes. Accordingly, the quality of HPP pasteurized food is very similar to that of fresh food products and the quality degradation is influenced more by subsequent storage and distribution rather than the pressure treatment. Pressure also provides unique opportunity to create and control novel food textures in protein based foods.

In some cases, pressure can be used to form protein gels and increase viscosity without using heat. HPP products currently marketed worldwide are primarily distributed refrigerated. In some cases this is necessary for safety (to prevent the growth of spores in low acid foods). For acid foods, refrigeration is not a necessity for microbial stability, but is employed to preserve flavour quality for extended periods of time.

HIGH-PRESSURE FOOD PROCESSING OF RICE AND STARCH FOODS

The use of pressure (P) in addition to temperature (T) has been accepted in the field of food science and technology over the past 15 years, and research and development are now under way in the food industry, universities, and government institutes. Several commercial products are now on the market that are prepared using high-pressure techniques. This section describes the principle of high-pressure treatment and the effects of high pressure on foods, with emphasis on the high-pressure effects on starches. Finally, recent successes of the Echigo Seika Company in the application of high-pressure techniques to rice and rice products are discussed.

PRINCIPLE AND METHOD

High pressure means high hydrostatic pressure generated by the compression of water. A pressure of 100 MPa or higher (usually lower than 1,000 MPa) is used under temperatures below 100 °C. A unit of pressure is expressed in kg cm^{-2}, bar, or pounds in^{-2}, but now the international Pascal unit is commonly used. Therefore, 1,000 bar or 1,000 kg cm^{-2} almost equals 100 MPa. Food, which is contained in a plastic bag and sealed by carefully removing air, is placed in a pressure vessel filled with water and high pressure is then generated in the vessel by a pressure pump.

Pressurization of Food

An egg is not crushed by compression at 600 MPa, but the egg white and yolk are coagulated. The colour of egg yolk of a pressurized egg is naturally yellow, whereas a boiled egg changes to a faded yellow. The colour difference is attributed to changes associated with pressurization: pressure affects only noncovalent bonding and coagulates proteins without splitting covalent bonds, thus keeping the colour and smell intact. High pressure induces protein denaturation in the same way as high temperature. Meat protein is also denatured by pressure treatment at 400 MPa, preserving the properties of raw meats. An example of prawns or shrimp is interesting: although a boiled prawn turns red and the meat coagulates, the appearance of a pressurized prawn is the same as that of raw shrimp, but the meat coagulates after pressurization at 400 MPa for 10 min.

High pressure also has an effect on starches: a thick suspension of rice starch forms a ball by standing against its own weight after pressurization at 700 MPa, indicating that starches are gelatinized by the pressure treatment.

Versatile Utility of High Pressure

A high-pressure treatment generally coagulates protein, thereby inactivating the enzymes, gelatinizing the starches, and killing microorganisms.

Thus, the use of high pressure promises to be a versatile process in food science and technology. Pressure produces a new texture in meats and starch-based foods, while keeping the original nutrients, flavour, and colour.

High-Pressure Effects on Pure Starches

High-pressure-induced Gelatinization of Starch

Effect of Pressure on Amylase Digestibility of Starches. The starches of potato, maize, and wheat are gelatinized by pressure treatment at warm conditions of 45–50°C. The pressurization produces unique properties that are different from those of heat-gelatinization: heat-treatment destroys starch granules, resulting in a transparent solution, but a pressure-treatment swells the granules while maintaining the granular structure. Nevertheless, amylolytic enzymes such as α-, β-, and glc-amylases digest the pressurized starches well, being similar to the phenomenon in which the pressure-treatment of proteins increases protease digestibility. The pressure-induced gelatinization of starches exhibits a sigmoid curve, suggesting that a two-state transition is involved, as in heat-induced gelatinization.

Effect of Pressurization Time on Amylase Digestibility of Starches. To attain full amylase digestibility of starches, pressurization under warm conditions for 2 to 6 h is necessary. Interestingly, pressurization of starches for a longer time makes amylase digestion difficult: the amylase digestibility of starches decreased by 20–50 per cent after pressurization for 17 h compared with the maximum digestibility obtained after pressurization for 2 to 6 h.

These observations suggest that pressure induces gelatinization of starches, similar to heating, but prolonged pressurization produces a new stable structure of starches, which is not susceptible to attack by amylase. *Birefringence of Starches after Pressurization.* The birefringence of starches is lost as increasing pressure is applied. Wheat starch is sensitive to pressure and birefringence is lost at 200 MPa.

In the pressurization of starches, the number of granules exhibiting complete birefringence decreases without increasing the incomplete birefringent granules. This loss of birefringence shows that the crystalline structure is destroyed by high pressure as well as by high temperature and that the high-pressure-induced gelatinization of starches follows a two-state transition without an intermediate state of destruction.

Physical Properties of Pressurized Starches

When a 50 per cent water suspension of starches is pressurized at 100 to 500 MPa and at 45°C for 1 h and air-dried, followed by analyses of their physical properties, the results are as follows. According to X-ray crystal analysis, the crystalline structure of potato, waxy maize, and maize starches decreases with an increase in pressure, although the crystalline structure of potato starch does not change up to 500 MPa.

The decrease in the high-pressure-induced crystalline structure is parallel with the increase in amylase susceptibility. Amylograms show that the transition temperature of pressurized starches is elevated, thus decreasing the viscosity. Interestingly, differential-scanning calorimetry (DSC) of pressurized starches shows that the peak temperature of the DSC patterns increases while the peak area decreases, indicating that some pressure-induced structural state of the pressurized starches is perturbed by low energy at higher temperatures.

Summary of Pressure-induced Changes in Starches

The structure of pressurized starches changes accompany-ing the increase in amylase digestibility, a loss of birefringence, and a loss of crystallinity. These unique structural properties, which are somewhat different from those of heat-gelatinized starches, should be analysed in detail for effective applications of high pressure to starch and related foods.

Rice-based Foods Produced by High-pressure Processing

Dr. Akira Yamazaki, of Echigo Seika Company Ltd., studied the properties of pressurized rice in detail to introduce the use of high pressure to the food industry. He equipped several large highpressure machines in his own factory and succeeded in sending rice and rice products to the market by introducing the high-pressure technique to the manufacturing process of rice products, cooked rice (*gohan*), rice crackers (*osembei*), and rice cakes (*omochi*).

High-pressure Effects on Rice Grains

Traditionally cooked-rice grains change their shape and develop some cracks, but rice grains after high-pressure pretreatment followed by heat-cooking swell, and their original shape is maintained without cracks.

High-pressure Cooked Rice for Microwave Ovens

Bread is purchased in the store and toasted just before eating. This is a typical life style, especially on a busy morning. However, 45 min are required to cook rice. Cooked rice tastes best and has its best texture just after steaming. Warming of cold cooked rice makes the taste worse. In general, heating a starch-based food twice leads to an unpalatable food. In history, instant-rice is a dream of Japanese consumers.

Dr. Yamazaki succeeded in producing oven-cooked rice with a good taste and texture for consumers, although the production scale is only enough to fulfill the requirement of a small city. He also succeeded in producing instant rice containing miscellaneous cereals. These instant cooked-rice cereals exhibited good taste and good texture by a 3-min heating in a microwave oven.

When high-pressure pretreated and successively heatcooked rice is compared with traditionally cooked rice, its properties are as follows: first, it is more gelatinized; second, it is more slowly retrograded; and third, it is gelatinized to a greater extent by heating just before serving. Japanese consumers who are particularly sensitive to rice accept these properties.

High-pressure Rice Cakes

During New Year days, the Japanese public is freed from the kitchen and enjoys the New Year cerebrations, eating rice cakes (*omochi*) every day, which are a preserved food. The company introduced the high-pressure technique for producing the traditional rice cake and a special kind of *omochi*, which contains herbs with a natural colour, smell, and taste, and this is now available on the market.

Merits of High-pressure Food Processing

High-pressure processing is useful not only for producing high-quality food, but also for improving the manufacturing process. The production time for rice crackers is shortened by introducing a high-pressure technique into the traditional processing system and the total energy cost is decreased by 10 per cent and the labour force by 56 per cent. The use of high pressure for food processing and cooking, in addition to heating and cooling, is now in our hands. Two factors, T and P, are useful for the manufacturing of good foods, including starch-based foods. Although T and P are used independently, their combined use is also important for the optimal use of high pressure. For example, heat-tolerant bacterial spores are inactivated by pressurization under elevated temperature.

INDUSTRIAL APPLICATION OF THE PROCESS FACES ENGINEERING PROBLEMS

So far only applications on a pilot plant scale are reported in the literature. For further developments on a larger scale, theoretical and practical problems should be solved.

The industrial application of the process faces engineering problems related to the movement of great volumes of concentrated sugar solutions and to equipment for continuous operations. The use of highly concentrated sugar solutions creates two major problems. The syrup's viscosity is so great that agitation is necessary in order to decrease the resistance to the mass transfer on the solution side. The difference in density between the solution (about 1.3 kg/litre) and fruit and vegetables (about 0.8 kg/litre), makes the product float. Another important aspect, so far not investigated, is the microbiological safety of the process, which should be studied thoroughly before further industrial development.

Osmoappertisation in the Processing of Apricots

In order to obtain an alternative to the canned fruit preserves and to maintain a high quality of the fruits, a research has been carried out on the osmoappertisation of apricots, a "combined" technique that consists in the appertisation of the osmodehydrated apricots. This technique could contributes also to the reduction of energy consumption, limits the cost of production and combines "convenience" (ready-to-eat, medium shelf-life) with many market outlets (retail, catering, bakery, confectionery, semi-finished products).

Apricot Processing

- *Fresh apricot puree:* After washing, cutting and removal of stones, apricot halves are dipped in 2% solution of sodium or potassium metabisulphite for 10 minutes. After draining, the resulting material is passed through a 0.045-in. screen pulper - finisher to produce a fresh apricot puree. The fresh apricot puree obtained in this way could be further processed in different semiprocessed (*i.e.* chemically or otherwise preserved products) or finished fruit products (fruit leathers, fruit bars, jams, etc.).
- *Concentrated apricot pulp:* Fresh apricot halves could also be steam blanched for 5 min., passed through a 0.045-in pulper - finisher and transformed in a purée with about 14 Brix depending on the fruit quality. This purée may be concentrated in steam jacketed kettles up to 20 Brix or in other adequate equipment (*e.g.* a stirred vacuum evaporator) up to 28°Brix.

  As for fresh apricot purée, the concentrate may be further

processed in various semiprocessed or finished fruit products as mentioned above and as will be described below.

- Dried apricot leather
  - From fresh fruit purée by drum drying. The fresh apricot purée at about 14 Brix could be dried to 12% moisture apricot sheets, using a double-drum dryer operating at 132 degrees C with a drum clearance of 0.008 in and speed of 45 sec per revolution.
  - From fruit concentrate by drum drying. The concentrate could also be dried to 12% moisture fruit sheets by the same process as described above.
  - From fresh fruit purée or from apricot concentrate by sun/solar drying or by dehydration.
- *Trays:* For sun/solar drying or dehydration of fruit pulp, the trays must have a solid base in order to retain the liquid contents. They may be made of metal, timber or plastic. Stainless steel or plastic trays are most suitable because they are unaffected by acid fruit pulp; they are, however, expensive.

  A metal tray could be 75 x 50 cm in size and with side 5 cm high. The trays must keep level during drying; if the tray is not level the pulp will run to the lowest point, giving a layer of irregular depth which will dry unevenly. Any tray which is not made of stainless steel or plastic must be covered inside with a sheet of heavy gauge plastic film to protect the pulp from chemical or bacteriological contamination.

  Standard sun/solar trays as described can be used by covering them inside with a sheet of plastic film to create a solid base.
- *Preparation before drying/dehydration*: Fresh apricot purée can be directly used for next processing steps. Fruit concentrate needs to be added to potassium metabisulphite to obtain a 0.3% concentration of SO2 in the material.
- *Drying/dehydration*: The apricot/fruit purée or concentrate is poured into the trays to a depth of about 1.5 cm. When stainless steel or plastic trays are used they should be coated with a thin layer of glycerine to prevent sticking.

The pulp is then sun/solar dried or tunnel/cabinet dehydrated; moisture content in the dried product should not exceed 14% and the SO_2 content should not be less than 1500 pp. The dried product is wrapped in cellophane to prevent sticking, then put inside polythene bags and stored at best in tight fitting tins and sealed to prevent moisture transfer.

From fresh fruit purée or from apricot concentrate, with sugar addition, and then processed by sun/solar drying or by dehydration. In some countries preference is for finished products with added sugar; and this is also interesting from a point of view of energy consumption (concentration is partially achieved by sugar dry matter) and of shelf life. The overall content in SO_2 could also be reduced as sugar is a preservation agent, the product will be close to a fruit "paste".

CONSERVATION OF MASS AND ENERGY

The law of conservation of mass states that mass can neither be created nor destroyed. Thus in a processing plant, the total mass of material entering the plant must equal the total mass of material leaving the plant, less any accumulation left in the plant. If there is no accumulation, then the simple rule holds that "what goes in must come out". Similarly all material entering a unit operation must in due course leave.

For example, if milk is being fed into a centrifuge to separate it into skim milk and cream, under the law of conservation of mass the total number of kilograms of material (milk) entering the centrifuge per minute must equal the total number of kilograms of material (skim milk and cream) that leave the centrifuge per minute.

Similarly, the law of conservation of mass applies to each component in the entering materials. For example, considering the butter fat in the milk entering the centrifuge, the weight of butter fat entering the centrifuge per minute must be equal to the weight of butter fat leaving the centrifuge per minute. A similar relationship will hold for the other components, proteins, milk sugars and so on.

The law of conservation of energy states that energy can neither be created nor destroyed. The total energy in the materials entering the processing plant, plus the energy added in the plant, must equal the total energy leaving the plant.

This is a more complex concept than the conservation of mass, as energy can take various forms such as kinetic energy, potential energy, heat energy, chemical energy, electrical energy and so on. During processing, some of these forms of energy can be converted from one to another. Mechanical energy in a fluid can be converted through friction into heat energy. Chemical energy in food is converted by the human body into mechanical energy.

For example, consider the pasteurizing process for milk, in which milk is pumped through a heat exchanger and is first heated and then cooled.

The energy can be considered either over the whole plant or only as it affects the milk. For total plant energy, the balance must include: the conversion in the pump of electrical energy to kinetic and heat energy, the kinetic and potential energies of the milk entering and leaving the plant and the various kinds of energy in the heating and cooling sections,as well as the exiting heat, kinetic and potential energies.

To the food technologist, the energies affecting the product are the most important. In the case of the pasteurizer, the energy affecting the product is the heat energy in the milk. Heat energy is added to the milk by the pump and by the hot water passing through the heat exchanger. Cooling water then removes part of the heat energy and some of the heat energy is also lost to the surroundings.

The heat energy leaving in the milk must equal the heat energy in the milk entering the pasteurizer plus or minus any heat added or taken away in the plant.

Heat energy leaving in milk = initial heat energy
+ heat energy added by pump
+ heat energy added in heating section
– heat energy taken out in cooling section
– heat energy lost to surroundings.

The law of conservation of energy can also apply to part of a process. For example, considering the heating section of the heat exchanger in the pasteurizer, the heat lost by the hot water must be equal to the sum of the heat gained by the milk and the heat lost from the heat exchanger to its surroundings.

From these laws of conservation of mass and energy, a balance sheet for materials and for energy can be drawn up at all times for a unit operation. These are called material balances and energy balances.

OVERALL VIEW OF AN ENGINEERING PROCESS

Using a material balance and an energy balance, a food engineering process can be viewed overall or as a series of units. Each unit is a unit operation. The unit operation can be represented by a box as shown in Figure. Into the box go the raw materials and energy, out of the box come the desired products, by-products, wastes and energy. The equipment within the box will enable the required changes to be made with as little waste of materials and energy as possible. In other words, the desired products are required to be maximized and the undesired by-products and wastes minimized. Control over the process is exercised by regulating the flow of energy, or of materials, or of both.

TYPES OF PROCESS SITUATIONS

CONTINUOUS PROCESSES

In continuous processes, time also enters into consideration and the balances are related to unit time. Thus in considering a continuous centrifuge separating whole milk into skim milk and cream, if the material holdup in the centrifuge is constant both in mass and in composition, then the quantities of the components entering and leaving in the different streams in unit time are constant and a mass balance can be written on this basis. Such an analysis assumes that the process is in a steady state, that is flows and quantities held up in vessels do not change with time.

Example: Materials Balance in Continuous Centrifuging of Milk

If 35,000 kg of whole milk containing 4% fat is to be separated in a 6 h period into skim milk with 0.45% fat and cream with 45% fat, what are the flow rates of the two output streams from a continuous centrifuge which accomplishes this separation? Basis 1 hour's flow of whole milk

Mass In

Total mass = kg.
Fat = 5833 × 0.04 = 233 kg.
And so water plus solids-not-fat = 5600 kg.

Mass Out

Let the mass of cream be × kg then its total fat content is $0.45x$. The mass of skim milk is $(5833 - x)$ and its total fat content is $0.0045(5833 - x)$.

Material balance on fat:

Fat in = Fat out

$5833 \times 0.04 = 0.0045(5833 - x) + 0.45x$.

and so $x = 465$ kg.

So that the flow of cream is 465 kg h^{-1} and skim milk (5833 – 465) = 5368 kg h^{-1}

The time unit has to be considered carefully in continuous processes as normally such processes operate continuously for only part of the total factory time. Usually there are three periods, start up, continuous processing (so-called steady state) and close down, and it is important to decide what material balance is being studied. Also the time interval over which any measurements are taken must be long enough to allow for any slight periodic or chance variation.

In some instances a reaction takes place and the material balances have to be adjusted accordingly.

Chemical changes can take place during a process, for example bacteria may be destroyed during heat processing, sugars may combine with amino acids, fats may be hydrolysed and these affect details of the material balance. The total mass of the system will remain the same but the constituent parts may change, for example in browning the sugars may reduce but browning compounds will increase. An example of the growth of microbial cells is given. Details of chemical and biological changes form a whole area for study in themselves, coming under the heading of unit processes or reaction technology.

Example: Materials Balance of Yeast Fermentation

Baker's yeast is to be grown in a continuous fermentation system using a fermenter volume of 20 m^3 in which the flow residence time is 16 h. A 2% inoculum containing 1.2 % of yeast cells is included in the growth medium.

This is then passed to the fermenter, in which the yeast grows with a steady doubling time of 2.9 h. The broth leaving the fermenter then passes to a continuous centrifuge which produces a yeast cream containing 7% of yeast, 97% of the total yeast in the broth. Calculate the rate of flow of the yeast cream and of the residual broth from the centrifuge.

The volume of the fermenter is 20 m^3 and the residence time in this is 16 h so the flow rate through the fermenter must be 20/16 = 1.250 $m^3\ h^{-1}$

Assuming the broth to have a density substantially equal to that of water, i.e. 1000 kg m^{-3},

Mass flow rate of broth = 1250 kg h^{-1}

Yeast concentration in the liquid flowing to the fermenter = (concentration in inoculum)/(dilution of inoculum)

= (1.2/100)/(100/2)

= 2.4×10^{-4} kg kg^{-1}.

Now the yeast mass doubles every 2.9 h, so in 2.9 h, 1 kg becomes 1×2^1 kg (1 generation).

In 16h there are 16/2.9 = 5.6 doubling times

1kg yeast grows to $1 \times 2^{5.6}$ kg = 48.5 kg.

Yeast in broth leaving = $48.5 \times 2.4 \times 10^{-4}$ kg kg^{-1}

Yeast leaving fermenter = initial concentration × growth × flow rate

= $2.4 \times 10^{-4} \times 48.5 \times 1250$ = 15 kg h^{-1}

Yeast-free broth flow leaving fermenter = (1250 – 15) = 1235 kg h^{-1}

From the centrifuge flows a (yeast rich) stream with 7% yeast, this being 97% of the total yeast:

The yeast rich stream is (15 × 0.97) × 100/7 = 208 kg h^{-1}

and the broth (yeast lean) stream is (1250 – 208) = 1042 kg h^{-1} which contains (15 × 0.03) = 0.45 kg h^{-1} yeast and the yeast concentration in the residual broth = 0.45/1042 = 0.043%

Table. Materials balance over the centrifuge per hour

Mass in	(kg)
Yeast-free broth	1235 kg
Yeast	15 kg
Total	1250 kg
Mass out	(kg)
Broth	1042 kg
(Yeast in broth	0.45 kg)

Yeast steam	208 kg
(Yeast in stream	14.55 kg)
Total	1250 kg

A materials balance, such as in Example for the manufacture of yeast, could be prepared in much greater detail if this were necessary and if the appropriate information were available. Not only broad constituents, such as the yeast, can be balanced as indicated but all the other constituents must also balance.

One constituent is the element carbon: this comes with the yeast inoculum in the medium, which must have a suitable fermentable carbon source, for example it might be sucrose in molasses.

The input carbon must then balance the output carbon, which will include the carbon in the outgoing yeast, carbon in the unused medium and also that which was converted to carbon dioxide and which came off as a gas or remained dissolved in the liquid.

Similarly all of the other elements such as nitrogen and phosphorus can be balanced out and calculation of the balance can be used to determine what inputs are necessary knowing the final yeast production that is required and the expected yields. While a formal solution can be set out in terms of a number of simultaneous equations, it can often be easier both to visualize and to calculate if the data are tabulated and calculation proceeds step by step gradually filling out the whole detail.

BLENDING

Another class of situations which arises includes blending problems in which various ingredients are combined in such proportions as to give a product of some desired composition. Complicated examples, in which an optimum or best achievable composition must be sought, need quite elaborate calculation methods, such as linear programming, but simple examples can be solved by straight-forward mass balances.

Example: Blending of Minced Meat

A processing plant is producing minced meat, which must contain 15% of fat. If this is to be made up from boneless cow beef with 23% of fat and from boneless bull beef with 5% of fat, what are the proportions in which these should be mixed?

Let the proportions be A of cow beef to B of bull beef.

Then by a mass balance on the fat,

Mass in Mass out

$A \times 0.23 + B \times 0.05 = (A + B) \times 0.15.$

that is $A(0.23 - 0.15) = B(0.15 - 0.05).$

$A(0.08) = B(0.10).$

$A/B = 10/8$

or $A/(A + B) = 10/18 = 5/9.$

i.e. 100 kg of product will have 55.6 kg of cow beef to 44.4 kg of bull beef.

It is possible to solve such a problem formally using algebraic equations and indeed all material balance problems are amenable to algebraic treatment. They reduce to sets of simultaneous equations and if the number of independent equations equals the number of unknowns the equations can be solved. For example, the blending problem above can be solved in this way.

If the weights of the constituents are A and B and proportions of fat are a, b blended to give C of composition c:

Then for fat $Aa + Bb = Cc$

and overall $A + B = C$

of which A and B are unknown, and say we require these to make up 100 kg of C then

$A + B = 100$

or $B = 100 - A$

and substituting into the first equation

$Aa + (100 - A)b = 100c$

or $A(a - b) = 100(c - b)$

or $A = 100\,(c - b)/\,(a - b)$

and taking the numbers from the example

$A =$

$= 55.6$ kg

and $B = 44.4$ kg

as before, but the algebraic solution has really added nothing beyond a formula which could be useful if a number of blending operations were under consideration.

RECONSTITUTION TEST FOR DRIED/DEHYDRATED PRODUCTS

In reconstitution water is added to the product which is restored to a condition similar to that when it was fresh. This enables the food product to be cooked as if the person was using fresh fruit or vegetable.

All vegetables are cooked but many of the dried fruits can be used for eating after they have been soaked in water. The following reconstitution test is used to find out the quality of the dried product.

Reconstitution test

1. Weigh out a sample of 35 grams from the bulked and packed final product of the previous day's production.
2. Put the sample into a small container (beaker) and add 275 ml of cold water (and 3.5 g salt).
3. Cover the container (with a watch-glass) and bring the water to the boil.
4. Boil GENTLY for 30 minutes.
5. Turn out the sample onto a white dish.
6. At least two people should then examine the sample for palatability, toughness, flavour and presence or absence of bad flavours. The testers should record their results independently.
7. The liquid left in the container should be examined for traces of sand/soil and other foreign matter.

This test can be used also to examine dried products after they have been stored for some time. Evaluation of rehydration ratio may be performed according to the following calculations.

Rehydration ratio. If weight of the dried sample is 10 g (Wd) and the weight of the sample after rehydration is 60 g (Wr), rehydration ratio is:

$$\frac{Wr}{Wd} = \frac{60}{10} = \frac{6}{1}, 6 \text{ to } 1$$

Rehydration coefficient. The weight of rehydrated sample is 60 g (Wr); the weight of dried sample is 10 g (Wd) and its moisture is 5% (Wu); raw material before drying had 87% water (A); rehydration coefficient is:

$$\frac{Wr}{\frac{Wd - Wu \times 100}{100 - A}}$$

$$= \frac{60 \times (100 - 87)}{10 - [10 \times 0.05]} = \frac{780}{9.5} = 52.1.$$

A simpler test for eating quality can be carried out without weighing and measuring. The material is placed in a cooking pot with water (and a little salt). The pot is then covered and boiled as described above. Except for a few products which are eaten in the dry state, most dried fruit and all dried vegetables are prepared by soaking and cooking. Often this preparation is carried out incorrectly and dried products get a bad reputation. Good quality dried products, after cooking and if properly treated should be similar to cooked fresh produce. In order to get good results, the following methods are recommended:

Quick Method

Cold water, ten times the weight of the dry product, is added to the dried product. The container is covered, brought to the boil and simmered GENTLY until the product is tender. The cooking time may be 15 to 45 minutes after the boiling point has been reached.

Slow Method

This gives better results than the quick method. Cold water is added to the dry food and is left to soak for 1 to 2 hours before cooking. The product is then cooked in the same water as that in which it was soaked. The actual cooking time will probably be shorter than that for the quick method.

Other points to remember are:

- If too much water is added the cooked product will have little flavour. However, if too little water is added the product may dry and burn. This can be avoided by adding small quantities of water during cooking;
- Always cook with a lid on the container;
- Salt, if required, should be added when the cooking is almost complete;

- Partly used packages of dry products should be reclosed tightly or kept in containers with good fitting lids.

THERMAL PROCESSING OF FOODS

Thermal processing implies the controlled use of heat to increase, or reduce depending on circumstances, the rates of reactions in foods.

A common example is the retorting of canned foods to effect sterilization. The object of sterilization is to destroy all microorganisms, that is, bacteria, yeasts and moulds, in the food material to prevent decomposition of the food, which makes it unattractive or inedible. Also, sterilization prevents any pathogenic (disease-producing) organisms from surviving and being eaten with the food. Pathogenic toxins may be produced during storage of the food if certain organisms are still viable. Microorganisms are destroyed by heat, but the amount of heating required for the killing of different organisms varies. Also, many bacteria can exist in two forms, the vegetative or growing form and the spore or dormant form. The spores are much harder to destroy by heat treatment than are the vegetative forms.

Studies of the microorganisms that occur in foods, have led to the selection of certain types of bacteria as indicator organisms. These are the most difficult to kill, in their spore forms, of the types of bacteria which are likely to be troublesome in foods.

A frequently used indicator organism is *Clostridium botulinum*. This particular organism is a very important food poisoning organism as it produces a deadly toxin and also its spores are amongst the most heat resistant. Processes for the heat treatment of foodstuffs are therefore examined with respect to the effect they would have on the spores of *C. botulinum*. If the heat process would not destroy this organism then it is not adequate. As *C. botulinum* is a very dangerous organism, a selected strain of a non-pathogenic organism of similar heat resistance is often used for testing purposes.

THERMAL DEATH TIME

It has been found that microorganisms, including *C. botulinum*, are destroyed by heat at rates which depend on the temperature, higher temperatures killing spores more quickly. At any given temperature, the spores are killed at different times, some spores being apparently more resistant to heat than other spores. If a graph is drawn, the number of surviving spores against time of holding at any chosen temperature, it is found experimentally that the number of surviving spores fall asymptotically to zero. Methods of handling process kinetics are well developed and if the standard methods are applied to such results, it is found that thermal death of microorganisms follows, for practical purposes, what is called a first-order process at a constant temperature.

This implies that the fractional destruction in any fixed time interval, is constant. It is thus not possible, in theory at least, to take the time when all of the organisms are actually destroyed. Instead it is practicable, and very useful, to consider the time needed for a particular fraction of the organisms to be killed.

The rates of destruction can in this way be related to:

- The numbers of viable organisms in the initial container or batch of containers.
- The number of viable organisms which can safely be allowed to survive.

Of course the surviving number must be small indeed, very much less than one, to ensure adequate safety. However, this concept, which includes the admissibility of survival numbers of much less than one per container, has been found to be very useful. From such considerations, the ratio of the initial to the final number of surviving organisms becomes the criterion that determines adequate treatment. A combination of historical reasons and extensive practical experience has led to this number being set, for *C. botulinum*, at 10^{12}: 1. For other organisms, and under other circumstances, it may well be different.

The results of experiments to determine the times needed to reduce actual spore counts from 10^{12} to 1 (the lower, open, circles) or to 0 (the upper, closed, circles) are shown in Figure.

In this graph, these times are plotted against the different temperatures and it shows that when the logarithms of these times are plotted against temperatures, the resulting graph is a straight line. The mean times on this graph are called thermal death times for the corresponding temperatures. Note that these thermal death times do not represent complete sterilization, but a mathematical concept which can be considered as effective sterilization, which is in fact a survival ratio of 1: 10^{12}, and which has been found adequate for safety.

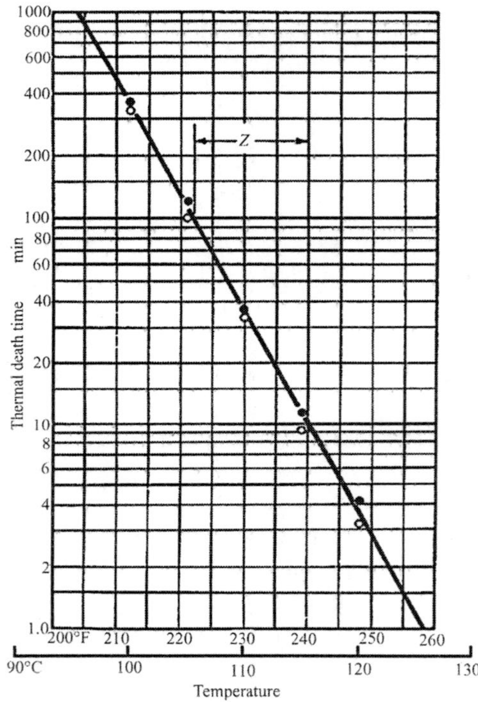

Fig. Thermal Death Time Curve for *Clostridium Botulinum*

Any canning process must be considered then from the standpoint of effective sterilization. This is done by combining the thermal death time data with the time-temperature relationships at the point in the can that heats slowest. Generally, this point is on the axis of the can and somewhere close to the geometric centre. Using either the unsteady-state heating curves or experimental measurements with a thermocouple at the slowest heating point in a can, the temperature-time graph for the can under the chosen conditions can be plotted. This curve has then to be evaluated in terms of its effectiveness in destroying *C. botulinum* or any other critical organism, such as thermophilic spore formers, which are important in industry. In this way the engineering data, which provides the temperatures within the container as the process is carried out, are combined with kinetic data to evaluate the effect of processing on the product. Considering Figure, the standard reference temperature is generally selected as 121.1°C (250 °F), and the relative time (in minutes) required to sterilize, effectively, any selected organism at 121°C is spoken of as the *F* value of that organism. In our example, reading from Figure, the *F* value is about 2.8 min. For any process that is different from a steady holding at 121°C, our standard process, the actual attained *F* values can be worked out by stepwise integration. If the total *F* value so found is below 2.8 min, then sterilization is not sufficient; if above 2.8 min, the heat treatment is more drastic than it needs to be.

EQUIVALENT KILLING POWER AT OTHER TEMPERATURES

The other factor that must be determined, so that the equivalent killing powers at temperatures different from 121°C can be evaluated, is the dependence of thermal death time on temperature. Experimentally, it has been found that if the logarithm of *t*, the thermal death time, is plotted against the temperature, a straight-line relationship is obtained. This is shown in Figure and more explicitly in Figure.

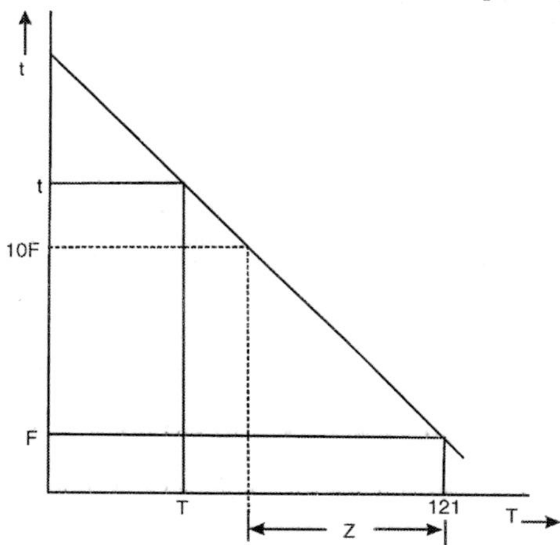

Fig. Thermal Death Time/Temperature Relationships

We can then write from the graph

$$\log t - \log F = m(121 - T) = \log t/F$$

where t is the thermal death time at temperature T, F is the thermal death time at temperature 121°C and m is the slope of the graph.

Also, if we define the z value as the number of degrees below 121°C at which t increases by a factor of 10, that is by one cycle on a logarithmic graph,

$t = 10F$ when $T = (121 - z)$

so that, $\log 10F - \log F = \log (10F/F) = 1 = m[121 - (121 - z)]$

and so $z = 1/m$

Therefore $\log (t/F) = (121 - T)/z$

or $t = F \times 10^{(121-T)/z}$

Now, the fraction of the process towards reaching thermal death, dS, accomplished in time dt is given by $(1/t_1)dt$, where t_1 is the thermal death time at temperature T_1, assuming that the destruction is additive.

That is $dS = (1/t_1)dt$

or $= (1/F)10^{-(121-T)/z}\, dt$

When the thermal death time has been reached, that is when effective sterilization has been achieved,

$$\int dS = 1$$

that is

$$\int (1/F)10^{-(121-T)/z} dt = 1$$

or

$$\int 10^{-(121-T)/z} dt = F$$

This implies that the sterilization process is complete, that the necessary fraction of the bacteria/spores has been destroyed, when the integral is equal to F. In this way, the factors F and z can be combined with the time-temperature curve and integrated to evaluate a sterilizing process.

The integral can be evaluated graphically or by stepwise numerical integration. In this latter case the contribution towards F of a period of t min at a temperature T is given by $t \times 10^{-(121-T)/z}$ Breaking up the temperature-time curve into t_1 min at T_1, t_2 mm at T_2, etc., the total F is given by

$F = t_1 \times 10^{-(121-T1)/z} + t_2 \times 10^{-(121-T2)/z} + ...$

This value of F is then compared with the standard value of F for the organism, for example 2.8 min for *C. botulinum* in our example, to decide whether the sterilizing procedure is adequate.

Example: Time/Temperature in a can during Sterilisation

In a retort, the temperatures in the slowest heating region of a can of food were measured and were found to be shown as in Figure. Is the retorting adequate, if F for the process is 2.8 min and z is 10°C?

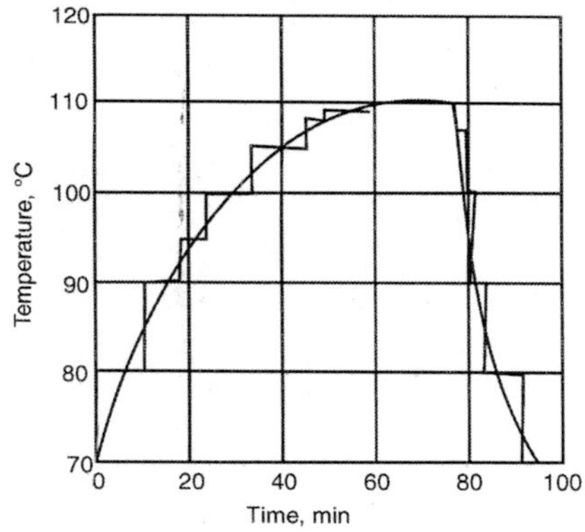

Fig. Time/Temperature Curve for can Processing

Approximate stepped temperature increments are drawn on the curve giving the equivalent holding times and temperatures as shown in Table. The corresponding F values are calculated for each temperature step.

Temperature T (°C)	Time t (min)	(121–7)	$10^{-1(121-T)10}$	$t \times 10^{-(121-T)10}$
80	11	41	7.9×10^{-5}	0.0087
90	8	31	7.9×10^{-4}	0.0063
95	6	26	2.5×10^{-3}	0.015
100	10	21	7.9×10^{-3}	0.079
105	12	16	2.5×10^{-2}	0.30
108	6	13	5.0 × 10–2	0.30
109	8	12	6.3×10^{-2}	0.50
110	17	11	7.9×10^{-2}	1.34
107	2	14	4.0×10^{-2}	0.08
100	2	21	7.9×10^{-3}	0.016

90	2	31	7.9×10^{-4}	0.0016
80	8	41	7.9×10^{-5}	0.0006
70	6	51	7.9×10^{-6}	0.00005

The above results show that the F value for the process = 2.64 so that the retorting time is not quite adequate. This could be corrected by a further 2 min at 110°C (and proceeding as above, this would add $2 \times 10^{-(121-110)/10} = 0.16$, to 2.64, making 2.8).

From the example, it may be seen that the very sharp decrease of thermal deadh times with higher temperatures means that holding times at the lower temperatures contribute little to the sterilization. Very long times at temperatures below 90°C would be needed to make any appreciable difference to F, and in fact it can often be the holding time at the highest temperature which virtually determines the F value of the whole process. Calculations can be shortened by neglecting those temperatures that make no significant contribution, although, in each case, both the number of steps taken and also their relative contributions should be checked to ensure accuracy in the overall integration.

It is possible to choose values of F and of z to suit specific requirements and organisms that may be suspected of giving trouble. The choice and specification of these is a whole subject in itself and will not be further discussed. From an engineering viewpoint a specification is set, as indicated above, with an F value and a z value, and then the process conditions are designed to accomplish this.

The discussion on sterilization is designed to show, in an elementary way, how heat-transfer calculations can be applied and not as a detailed treatment of the topic.

PASTEURIZATION

Pasteurization is a heat treatment applied to foods, which is less drastic than sterilization, but which is sufficient to inactivate particular disease-producing organisms of importance in a specific foodstuff. Pasteurization inactivates most viable vegetative forms of microorganisms but not heat-resistant spores. Originally, pasteurization was evolved to inactivate bovine tuberculosis in milk. Numbers of viable organisms are reduced by ratios of the order of 10^{15}: 1. As well as the application to inactivate bacteria, pasteurization may be considered in relation to enzymes present in the food, which can be inactivated by heat. The same general relationships as were discussed under sterilization apply to pasteurization. A combination of temperature and time must be used that is sufficient to inactivate the particular species of bacteria or enzyme under consideration. Fortunately, most of the pathogenic organisms, which can be transmitted from food to the person who eats it, are not very resistant to heat. The most common application is pasteurization of liquid milk. In the case of milk, the pathogenic organism that is of classical importance is *Mycobacterium tuberculosis*, and the time/temperature curve for the inactivation of this bacillus is shown in Figure.

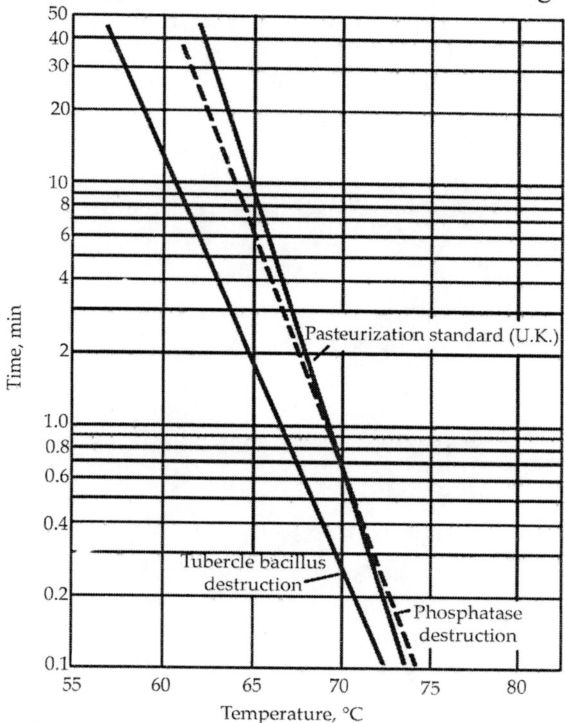

Fig. Pasteurization Curves for Milk

This curve can be applied to determine the necessary holding time and temperature in the same way as with the sterilization thermal death curves. However, the times involved are very much shorter, and controlled rapid heating in continuous heat exchangers simplifies the calculations so that only the holding period is really important.

For example, 30 min at 62.8°C in the older pasteurizing plants and 15 sec at 71.7°C in the so-called high temperature/short time (HTST) process are sufficient. An even faster process using a temperature of 126.7°C for 4 sec is claimed to be sufficient. The most generally used equipment is the plate heat exchanger and rates of heat transfer to accomplish this pasteurization can be calculated by the methods explained previously. An enzyme present in milk, phosphatase, is destroyed under somewhat the same time-temperature conditions as the *M. tuberculosis* and, since chemical tests for the enzyme can be carried out simply, its presence is used as an indicator of inadequate heat treatment. In this case, the presence or absence of phosphatase is of no significance so far as the storage properties or suitability for human consumption are concerned.

Enzymes are of importance in deterioration processes of fruit juices, fruits and vegetables. If time-temperature relationships, such as those that are shown in Figure for phosphatase, can be determined for these enzymes, heat processes to destroy them can be designed. Most often this is done by steam heating, indirectly for fruit juices and directly for vegetables when the process is known as blanching.

The processes for sterilization and pasteurization illustrate very well the application of heat transfer as a unit operation in food processing. The temperatures and times required are determined and then the heat transfer equipment is designed using the equations developed for heat-transfer operations.

Example: Pasteurisation of Milk

A pasteurization heating process for milk was found, taking measurements and times, to consist essentially of three heating stages being 2 min at 64°C, 3 min at 65°C and 2 min at 66°C. Does this process meet the standard pasteurization requirements for the milk, as indicated in Figure, and if not what adjustment needs to be made to the period of holding at 66°C?

From Figure, pasteurization times t_T can be read off the UK pasteurisation standard, and from and these and the given times, rates and fractional extents of pasteurization can be calculated:

At 64°C, $t_{64} = 15.7$ min

so 2 min is $\dfrac{2}{15.7} = 0.13$

At 65°C, t_{65} = 9.2 min

so 3 min is $\frac{3}{9.2} = 0.33$

At 66°C, t_{66} = 5.4 min

so 2 min is $\frac{2}{5.4} = 0.37$

Total pasteurization extent = (0.13 + 0.33 + 0.37) = 0.83.
Pasteurization remaining to be accomplished = (1 – 0.83) = 0.17.
At 66°C this would be obtained from (0.17 × 5.4) min holding = 0.92 min.

So an additional 0.92 min (or approximately 1 min) at 66°C would be needed to meet the specification.

HEATING COILS IMMERSED IN LIQUIDS

In some food processes, quick heating is required in the pan, for example, in the boiling of jam. In this case, a helical coil may be fitted inside the pan and steam admitted to the coil as shown in Figure. This can give greater heat transfer rates than jacketed pans, because there can be a greater heat transfer surface and also the heat transfer coefficients are higher for coils than for the pan walls.

Examples of the overall heat transfer coefficient *U* are quoted as:

- 300-1400 for sugar and molasses solutions heated with steam using a copper coil,
- 1800 for milk in a coil heated with water outside,
- 3600 for a boiling aqueous solution heated with steam in the coil.

with the units in these coefficients being J m^{-2} s^{-1} °C^{-1}.

SCRAPED SURFACE HEAT EXCHANGERS

One type of heat exchanger, that finds considerable use in the food processing industry particularly for products of higher viscosity, consists of a jacketed cylinder with an internal cylinder concentric to the first and fitted with scraper blades, as illustrated in Figure. The blades rotate, causing the fluid to flow through the annular space between the cylinders with the outer heat transfer surface constantly scraped. Coefficients of heat transfer vary with speeds of rotation but they are of the order of 900-4000 J m^{-2} s^{-1} °C^{-1}. These machines are used in the freezing of ice cream and in the cooling of fats during margarine manufacture.

PLATE HEAT EXCHANGERS

A popular heat exchanger for fluids of low viscosity, such as milk, is the plate heat exchanger, where heating and cooling fluids flow through alternate tortuous passages between vertical plates as illustrated in Figure. The plates are clamped together, separated by spacing gaskets, and the heating and cooling fluids are arranged so that they flow between alternate plates. Suitable gaskets and channels control the flow and allow parallel or counter current flow in any desired number of passes. A substantial advantage of this type of heat exchanger is that it offers a large transfer surface that is readily accessible for cleaning. The banks of plates are arranged so that they may be taken apart easily. Overall heat transfer coefficients are of the order of 2400-6000 J m^{-2} s^{-1} $°C^{-1}$.

DAIRY PROCESSING INDUSTRY

The dairy industry has high losses ranging from 2 per cent to 3 per cent volume of milk intake (80 to 130 megalitres annually). Depending on the particular site configuration and the waste minimization processes employed, the volume of wastewater generated during dairy processing may be as high as 2½ litres of wastewater per litre of milk processed. (World's best practice is 0.5 litres of wastewater per litre of milk processed.) Poorly treated wastewater with high levels of pollutants—caused by poor design, operation or treatment systems—creates major environmental problems when discharged to surface water or land. Such problems include:

- Contamination and deoxygenation of streams and waterways by direct discharge or run-off of inadequately treated wastewater.
- Excessive concentration of nutrients such as nitrogen and phosphorus in surface and subsurface waters (this contributes to excessive growth of plants and algae blooms which makes downstream water unsuitable for domestic, agricultural and industrial uses).
- Land degradation and damage to pastures and crops-long-term damage to soil productivity may arise from:

- excessive nutrient loading
- High salinity
- Low/high pH
- over-application of wastewater to land, resulting in contaminated groundwater
- Soil structure decline due to wastewater with high sodium adsorption ratio
- Poor irrigation design
- clogging of soils by fats/solids from irrigated wastewater (this may lead to restricted infiltration that will promote pooling and subsequent odour).

If wastewater supply exceeds the soil's evapouration and infiltration capacity, the problems increase. Offensive odours from wastewater being stored or treated are often a problem—particularly with such strong wastes as whey.

Extended storage periods are more likely in the cooler months when irrigation is not practicable. Water and land or reuse systems should be monitored regularly because of the high chance of detrimental environmental effects. The monitoring should be followed by performance assessment and remedial action if necessary.

Environmental Policy and Guidelines

Waste management and reuse facilities must comply with the relevant environmental policies including those listed below:

- *State Environment Protection Policy (the Air Environment)*, which specifies the objectives for specific gaseous components and particulate emissions.
- *State Environment Protection Policy (Waters of Victoria)*: Schedule E lists emission limits for waste discharges to water. More stringent requirements may be necessary with environmental sensitive water areas.
- *State Environment Protection Policy (Waters of Victoria)*: Clause 22 requires that waste be discharged preferably to land where practical and environ-mental beneficial.
- Wastewater disposal facilities should be designed in accordance with *Guidelines for Wastewater Irrigation* (EPA Publication 168). In particular, wastewater storage and disposal should be designed and built to contain all waste

in at least 90 per cent of wet years. People and organizations that use land irrigation should ensure that there is enough land area for both present and future wastewater disposal.

- *State Environment Protection Policy (Waters of Victoria)*: Clause 23 refers to minimization of waste generation.
- State Environment Protection Policy (Groundwaters of Victoria) (draft 1994) requires that groundwater is protected from activities potentially detrimental to its quality, as well as from hydrogeological testing and assessment.
- State Environment Protection Policy.
- Industrial Waste Management Policy (Waste Minimization). 1990 requires all premises subject to works approval to have comprehensive waste management plans—covering all aspects of waste minimization and identifying options for waste minimization, the handling, storage and disposal of wastes. These plans are to be prepared in accordance with EPA guidelines.

The aim of these guidelines is to provide the dairy industry with:

- A clear statement of environmental objectives for each segment of the environment affected.
- Methods to assess and minimize the actual or likely impacts of the dairy industry.
- On the basis of available experience, an outline of BPEM practices to achieve the desired results.

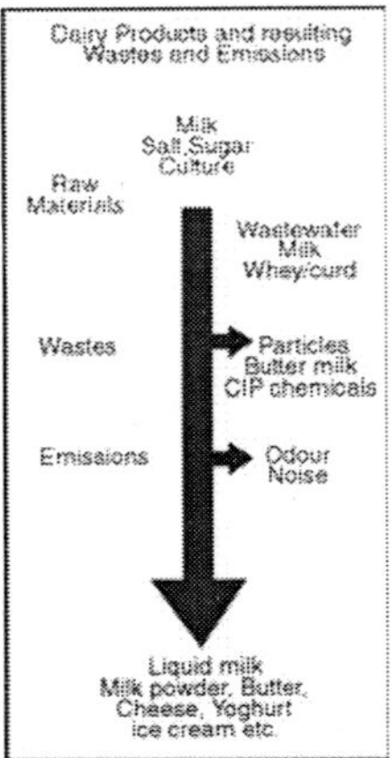

Fig. Dairy Wastes and Emissions

Cleaner Production

Waste Minimization

Under the Industrial Waste Management Policy (Waste Minimization), 1990, premises which are subject to works approval require waste management plans incorporating waste minimization.

Each dairy plant should therefore assess opportunities for reducing waste arising from its operations. Waste reduction measures may include:

- Reducing use of water.
- Reducing use of chemicals or substitution of mineral salts, for example, potassium in place of sodium compounds.
- Recycling water and chemicals.
- Recovery and reuse of product from first reuse.
- Reuse/reprocessing of off-spec material.
- Recovering and reusing spilled raw materials and products.

Process Control

Dairy configuration and the products made affect the nature and concentration of dairy wastes. The amount of product lost depends on design and operational factors including:

- The range of process technologies in use.
- The availability of adequate process monitoring, and plant and procedure alarms/interlocks.
- The availability of automated operation—especially automated clean-in-place (CIP) systems and procedures.
- The level of management and operator commitment, training and efficiency.
- The level of routine equipment maintenance.

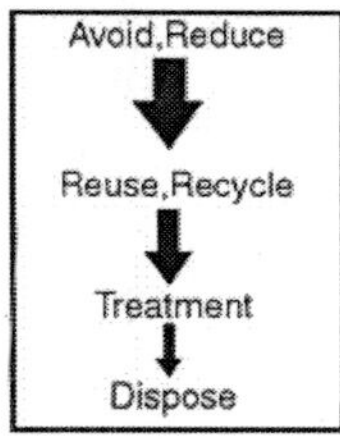

Fig. Best Practice Management Options for Waste in Order of Preference

Most site losses come from activities associated with liquid handling and, to a lesser extent, with the discharge of air and solid waste. Some examples of avoidable losses are:

- Leaking valves, pumps, pipelines or other fittings—the volume lost may not be large but the pollution load may be great.
- Spills from overflows, malfunctions and poor handling procedures—spills usually happen over a short period but the amount and the high concentra-tion of milk or product lost may be a significant increase in the pollution load.
- Losses from processing and cleaning during the normal operation of plant and equipment—this includes the deliberate discharge of unwanted materials such as whey, spent cleaners and diluted product not thought to be worth recovering.

Suggested Measures for Reducing Waste

Butter and Dried Products

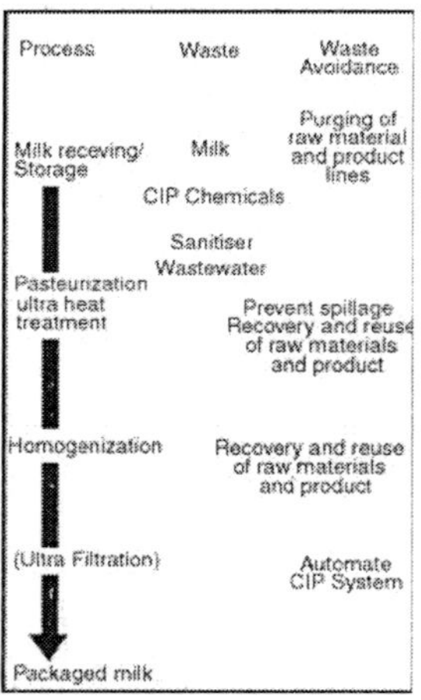

Fig. Avoiding Waste During Liquid Milk Production

Best practice involves processing the predominant by-products such as whey, buttermilk and skim milk, into high value products like skim milk powder (SMP), buttermilk powder (BMP), whey powder, whey protein concentrate and casein, rather than being used as low value animal feed/fertilizer or being dumped as waste. Cream and butter are viscous and fatty and stick to equipment surfaces much more strongly than liquid milk, increasing the problem of removing residues. Hot water is an effective way to remove residual butterfat from cream processing and butter making equipment but the water temperature must not be too high (< 65°C) or there is the risk of "burning on" of some of the proteins. Whey should be dried where possible.

Production Performance

To initiate the process of cleaner production, a change in culture to waste minimization is required. This involves moving from pollution treatment and control to anticipation and prevention of wastes. Ways to prevent the build up of surface deposits include:

- Minimization of surface area.
- Prevention of build-up of milkstone deposits.
- Maintenance of butter churns.
- Correct preparation before filling.
- Not over-working the batch (this does not often happen with continuous butter making).

To avoid spills, buttermilk collection facilities should be large enough to hold all buttermilk discharged. Buttermilk should be dried or used as animal feed and solids recovered from butter wash water also may be sold as stock feed.

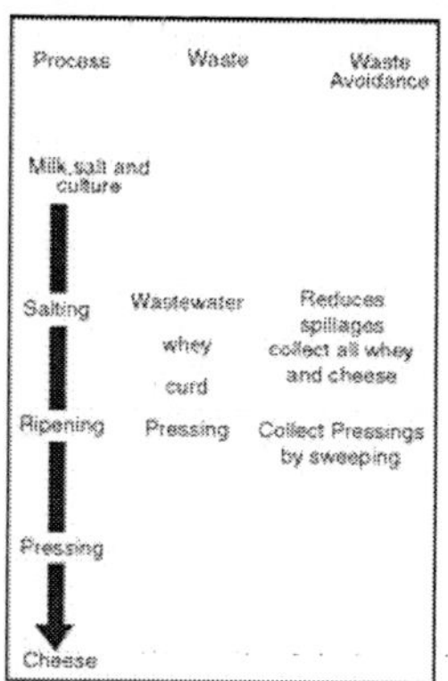

Fig. Ways to Avoid Waste During Butter Production

Cheese and Dried Products

Making cheese generates a large volume of by-products such as whey. Waste reduction can be achieved by:

- Not overfilling cheese vats to stop curd loss.
- Completely removing whey and curds from vats before rinsing.
- segregating all whey drained from cheese.
- Sweeping up pressings (particles).
- Screening all liquid streams to collect fines.

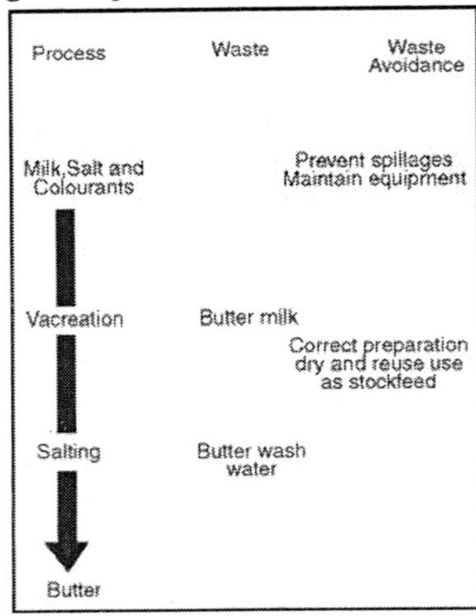

Fig. Ways to Avoid Waste during Cheese Production

Evapouration and Powder Production

It is suggested that evapourators be operated to:

- Maintain a liquid level low enough to stop product boil-over.
- Run to specified length—excessively long runs with higher than specified running rates lead to blocked tubes which not only produce high pollution, but are difficult and time-consuming to clean.
- Use effluent entrainment separators to avoid carry-over of milk droplets during condensation of evapourated water.
- Recirculate low concentration milk or other feed-stock until it reaches the required concentration.
- Process rinsings with 7 per cent or more of solids before scheduled shutdowns, or evapourate them during the next run rather than discharging to the sewer.

- Minimize air emissions by using fabric filters or wet scrubbers.

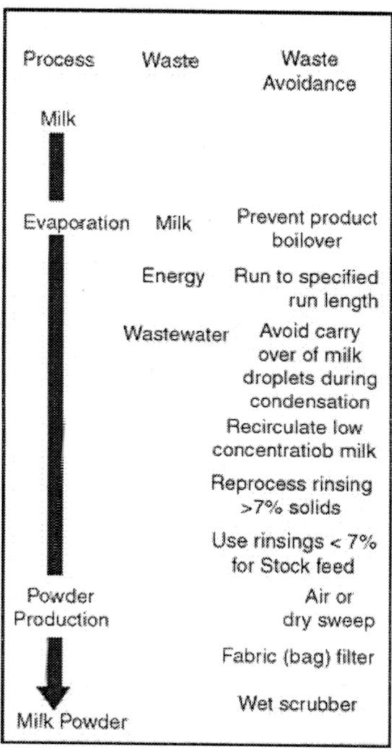

Fig. Ways to Avoid Waste During Powder Production

Environmental Elements

The selection of a site for the construction, replacement or expansion of a dairy plant should take into consideration nearby land uses, possible future developments, the volumes and nature of wastes produced and the proposed nature of waste recycling, reuse or disposal. Depending on the proposed waste disposal system, adequate land should be available for treatment of wastewater. Soil types should also be assessed on site to check whether they can provide reasonable drainage and have a good capacity to retain nitrogen, phosphorus and organic matter.

Generally, soils with textures ranging from medium loams to medium clays are suitable. Sandy soils are not suitable because of the risk of leaching of contaminants into underlying groundwater.

Similarly, wastewater should not be applied to heavy clay soils where water logging or surface run-off may occur. Dairy plants and their associated wastewater treatment plants should not be located on a flood plain and should be a sufficient distance from surface water bodies and wetlands to reduce the risks of contamination caused by run-off or accidental spills. Similarly, wastewater treatment and disposal areas should not be sited above major ground water recharge areas such as gravel or sand beds or fractured rock aquifers.

Buffer Distances

In order to provide a basic level of protection from odour, dust and noise, a dairy plant should not be located within a minimum buffer distance of designated residential areas or other sensitive land uses. This is to protect the amenity of the area from unintended or accidental emissions which may arise from causes such as equipment failure, accidents and abnormal weather conditions. The buffer distance is measured from the nearest dairy activity capable of emitting odour or particle emissions.

Air modelling studies may be necessary at the design stage for large operations when the buffer distances are close to the recommended minima. Siting should also consider the need to protect sensitive natural water resources. Thus a dairy plant should not be sited within 100 metres of surface waters, nor be located on a flood plain or in declared special water supply catchment areas—unless adequate protection of surface and groundwaters can be demonstrated by the proponent. An ideal buffer distance between dairy processing operations and residential areas would be at least a kilometre. However, for facilities that operate their own anaerobic, aerobic or facultative wastewater treatment lagoons and irrigation-based disposal areas, this separation distance may vary from 200 to 2,200 metres—the distance depends on the flow rate and strength of the treated and disposed wastewater.

Site Selection and Siting

Buffer distances usually are confined to new or *greenfield* sites, or to additional work on existing sites. If there is to be substantial development on existing sites with inadequate buffer zones, the site developers usually must show that the introduced technology will allow for a variation of the recommended buffer distance. If the recommended buffer distance is to be reduced for site-specific circumstances, the following performance criteria should be met:

- The plant's standard emission control technology is significantly better than the good level assumed by the buffer distance guidelines. Normally best practice technology is the necessary standard.
- An environmental audit of the site emissions has been undertaken.
- There has been no recent history of complaints from residual emissions for an existing plant.
- The topographic or meteorological characteristics which affect dispersion of emissions are evaluated. The following requirements may be taken into account for minimising off-site emissions for any given site with defined boundaries:
- Locate emission points to give maximum separation distances from adjoining residential or other sensitive areas with consideration of prevailing winds (Ausplume).
- Use clean technology and/or best practice control technology.
- Ensure there are adequate operating practices, housekeeping and liaison with the local community.

Types of Emissions

The main emissions from dairy manufacturing processes are odours and particles.

Odours

Odours in and around milk processing plants come from the biological decomposition of milkderived organic matter, generally found in wastewater. Often these odours are due to poor housekeeping, overloaded or improperly run wastewater treatment and disposal facilities, and prolonged storage of strong wastes such as whey.

Particles

Particle emissions are caused either by combustion of solid or liquid fuel or, more often, spray drying of milk and whey. Excessive emissions are often sporadic and happen during plant upsets, shutdowns or startups. The use of solid or liquid fuel such as briquettes and oil can result in fallout—carbonaceous ash particulate is usually emitted during boiler upsets or tube soot-blowing operations. Milk powder particles—while not toxic—accumulate on flat surfaces such as roofing, guttering and rainwater tanks, and may seriously compromise the quality of stormwater discharged from the site or taint the drinking water.

A further source of annoyance to residents and factory workers is powder settling on nearby motor vehicles. The drier emissions depend on the product being dried—for example, skim milk tends to result in the highest emissions.

Source of Noise

The State Environment Protection Policy requires that any noise due to activities at a premises in a sensitive area must not exceed the noise limits for the area, as determined by the methods set out in the policy.

The limits are tighter outside the normal working day—such as the evening and especially at night. Most milk processing plants are located in country areas where there are no residential statutory requirements for industry but EPA may set noise targets based on the state environment protection policy and guidelines and may use discretion in each particular case.

In addition, there may be local government requirements on industrial noise. The principal causes of continuous noise include:

- Air discharges from drier stacks.
- Heater fans.
- Air supply fans.
- Ventilation.
- Boilers.
- Pumps.
- Cooling towers.
- Refrigeration units.
- Aerators on aerated lagoons.

Truck movements to and from the site or in streets are a source of noise, as are refrigeration compressors on trucks. This is a particular problem when fresh milk delivery means late night trucking.

Noisy operations at dairy plants include milk drying—which requires high airflows—and the movements of transport vehicles to and from the site.

Depending on the distance to sensitive receptors such as residential areas, suitable noise suppression or abatement measures—such as noise silencers on equipment, enclosure of outdoor equipment, concrete housing for mechanical plant, mufflers on transport vehicles—may be required. To reduce noise from aerated lagoons, adopt minimum capacity.

It may also be necessary to avoid carrying out certain operations or reduce noise levels:

- Before 7 am and after 6 pm on weekdays.
- Before 7 am and after 1 pm on Saturday.
- On Sunday.
- On public holidays.

A detailed professional assessment of noise emissions should also be conducted.

Wastewater Quality

Protection of Surface and Ground waters

Activities at dairy plants have the potential to contaminate both surface waters and groundwater. Water and land pollution can be avoided by appropriate siting, design, management and control of the dairy plant.

Sources of Dairy Wastewater

Approximately 65 per cent of dairy factory losses enter wastewater discharge streams and these can have a major impact on the environment. The main sources of dairy processing plant wastewater are:

- Raw material (predominantly milk) and product losses from leaking equipment and pipelines, and spills caused by equipment overflows and malfunctions and by poor handling procedures.
- Materials used for cleaning and sanitizing.
- By-products such as whey from the manufacture of cheese and casein.

Components of Dairy Wastewater

The major contaminants in dairy processing wastewater are milk solids that contain milk fat, protein, lactose and lactic acid. Other minor constituents include sodium, potassium, calcium and chloride. Organic wastewater strength is measured by either BOD or COD. Typical process wastewater has a biochemical oxygen demand (BOD) of about 2,000 mg/L and a dissolved solids concentration of 1,800 mg/L. BOD_5 is a measure of the amount of organic matter that is able to be biologically oxidised over a five day period.

Whey has a BOD concentration of 30,000- 40,000 mg/L. Where the whey is not used as a by-product but is discharged as effluent, it will increase the BOD level of wastewater and cause treatment and disposal problems. Whole milk has a BOD of 100,000 mg/L. Although the throughput of milk in dairy plants is generally increasing, the technologies available for reducing and recycling wastes means that the volume of water used and wastewater generated is significantly less in modern plants.

Wastewater Treatment and Disposal

Because of the highly seasonal nature of milk p: oduction, during peak periods the volume of wastewater generated at dairy plants may be several times greater than during off peak periods. The batch nature of many processes, and intermittent operations such as cleaning and sanitising, also means a wide daily variation in wastewater flows and quality. Options for dairy factory wastewater include:

- Treatment to a suitable standard for reuse or recycling.
- Discharge to local authority sewers under a trade waste agreement (with pre-treatment as necessary).
- Appropriate treatment and land discharge wherever practicable and environmentally beneficial.

Reuse and Recycling

Many dairy plants have technologies in place for recovering wastewater and/or condensate (from production of milk powder) for reuse in the dairy plant. Reuse and recycling can considerably decrease the volume of mains water required to operate the plant and also reduce the cost of both mains water and wastewater disposal. Fats, milk solids and minerals can also be recovered from wastewater and recycled—either at the dairy plant or offsite. Cleaning chemicals can also be recovered and reused on site.

Treatment and Discharge to Land

Dissolved salts contained in dairy plant wastewater can adversely affect soil structure if wastewater is used to irrigate land. Wastewater can also leach into underlying groundwater and affect its quality. Dairy plants should maximize the recovery, recycling and reuse of acids and alkali to minimise the dissolved salts and sodium levels in the wastewater. High salt levels affect the type of vegetation that grow. Over-irrigation may cause the underlying water table to rise, resulting in further deterioration of surface soils and vegetation.

Discharge to Sewer

Even if there is a local sewage authority that will accept wastewater, the volume and organic load of wastewater from just one dairy factory during peak season may well exceed the township's domestic waste. This may overload the sewage treatment plant, cause odours and give rise to poor effluent quality. Domestic wastewaters have a BOD_5 concentration of about 250 to 300 mg/L but in peak season a large dairy factory could be discharging two megalitres of wastewater at BOD_5 of 2,000 mg/L each day—the additional load on a sewerage plant is equivalent to an extra 16,000 persons.

COMPONENTS OF A WASTE WATER SYSTEM

SEGREGATION

Clean stormwater should be separated from contaminated stormwater and discharged directly into stormwater drains. Waste streams from the plant should also be segregated—for example, whey can be reused to produce whey powder or stockfeed. Spent cleaning solutions should be separated from other wastewater streams as they can be treated to recover cleaning agents. Highly saline wastewater should also be discharged separately to an evapouration pond where the salts can be recovered and recycled.

Equalization and pH Control

A balance tank or pond will even out variations such as pH and temperature. Alternative pH control can be achieved by using spent acid and alkali cleaners to neutralise each other.

Fat Removal

Coarse milk solids should be removed by screening. Fats can constitute up to 50 per cent of the organic load. Its recovery is therefore significant in any treatment process.

Dissolved air flotation is a very effective method of separation.

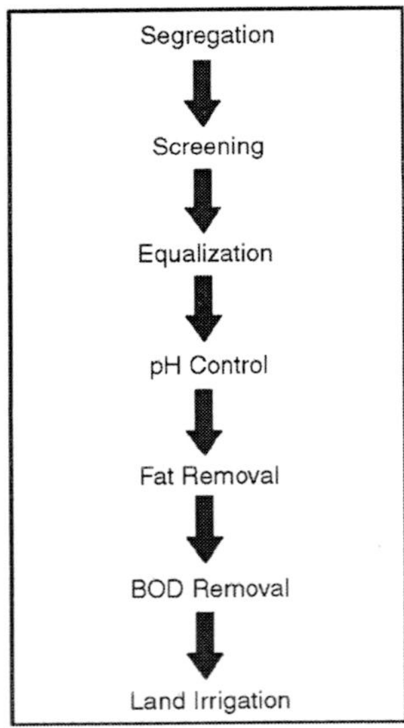

Fig. Best Practice for Waste Water Systems

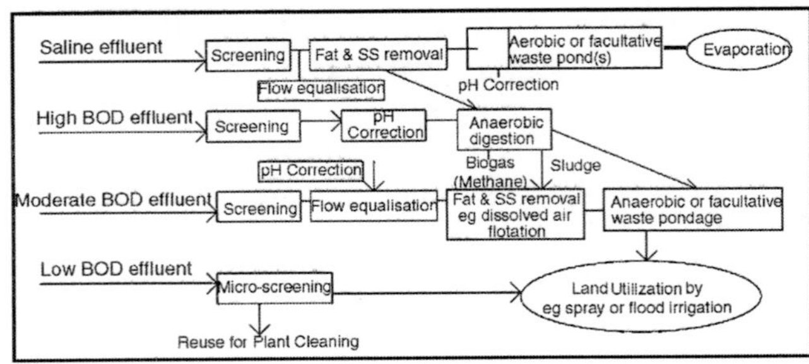

Fig. Suggested Options for Treatment of Segregated Waste Streams

Removal of Organic Load

Organic load can be reduced by physical methods—such as microfiltration, reverse osmosis and flotation techniques—or by biological treatments—such as activated sludge systems, trickling filters and anaerobic digesters. Lagoons, land irrigation and grass filtration systems can also reduce organic loads but reduction will occur at a slower rate than the previous methods. Best practice management of the waste stream may include removal of product before treatment.

Biological Processes

Activated Sludge

A highly effective method for treatment of dairy plant wastewater is the oxidation ditch. This is a development of the extended aeration process where aeration, settling and withdrawal of effluent all takes place in the same tank. The oxidation ditch process is characterised by a long retention time and low net sludge yield. This type of treatment lends itself to biological nitrogen removal.

Trickling Filters

The best trickling filters have a free passage of air to prevent the generation of odours but are sensitive to high or low pH which may result in killing the biomass.

Lagoons

Highest quality wastewater and low odour generation can be achieved in aerated lagoons which use floating aerators to force oxygen input and resemble activated sludge systems.

Advanced Treatment for Reuse

Membrane Filtration

This process has the potential for acid and alkali recovery and recycling. Best quality wastewater is obtained by pumping effluent through porous media containing millions of tiny pores. The media area is regularly cleaned by high pressure backwash using water and/ or air.

Reverse Osmosis

The removal of dissolved solids is best achieved by the passage of water through a semi-permeable membrane that restricts the movement of salts. This process for the desalination of wastewater is based on the osmotic pressures on either side of the membrane.

Land Irrigation

In inland areas, treated wastewater can be either sprayed on the land or applied by flood irrigation using laser levelling and a ridge and furrow system. Grass filtration allows a greater application rate than irrigation and grass filtration systems generally are intended as a refining phase in the overall treatment system. In these cases, the majority of the BOD will have been removed by other means.

Emergency Storage

High strength waste spills can be directed to small dams which have been built to prevent the treatment process from turning anaerobic.

Solid Wastes

Solid waste often is the easiest to see, quantify and correct, and the volumes, treatment, transport and disposal costs are likely to be very evident on a plant's balance sheet. Solid waste is generally:

- Off-spec product—for example, milk powders and heavy consistency product.
- Defective product packaging—for example, paperboard cartons, plastic containers.
- Recovered wastewater treatment sludges.
- Solid and semi-solid intermediate or finished product spills.

Minimization strategies range from commonsense housekeeping to detailed production planning to eliminate the generation of solid waste—for example:

- Train staff so that spillages are minimized
- Sweep up spills of solid materials such as cheese curd and powders to use as stock food—they should not be washed down the drain as this means more water is used and the volume and loading of wastewater is increased.
- Fit drain openings with screens and/or traps so that solid

waste does not enter the wastewater treatment/disposal system—these solid interception devices must be maintained regularly, preferably daily.

- Use "off-spec" or contaminated product as stock food—if it is disposed to land, it must be in accordance with EPA requirements.

Plant Management

Careful attention to quality assurance of the product is essential not only for marketing but also for responsible environmental management. To achieve and sustain world best environmental management practices it is recommended that:

- A quality management system and an environmental management system that conform with the principles and practices of ISO9001 and ISO14001 be adopted and implemented.
- An effective audit programme be prepared and implemented to ensure that all resources have been effectively utilized.
- An effective training programme be prepared and implemented for all employees involved in control and verification activities.
- A total quality management approach is adopted for the purpose of maintaining continuous improvement in process development through problem solving.
- A protocol for quality assurance/quality control—particularly in relation to sampling and analysis—be adopted.

To achieve a consistently high level of environmental performance, a sound environmental management system is essential. The Environment Management System (EMS) Standard ISO 14001 provides a useful guide. BPEM for dairy plants includes:

- A commitment from management to an environmental policy which is communicated to all employees.
- Adherence to BPEM guidelines.
- Alert and informed supervision.
- Regular operator training.
- Exercising control over the product and the process through a QA/QC protocol.

- Detailed written procedures for each activity which are used by operational staff.
- Understanding and control of all wastes and emissions.
- Contingency plans, including:
  - Procedures consistent with these guidelines for treating wastewater.
  - Procedures to minimise emissions (odour, particulates) to the air environment.
  - Procedures to minimise the generation of noise.
- A high level of housekeeping on the site.
- Continuous improvement.

A well managed dairy plant would be expected to have an open attitude and good relations with the community, and a reputation of being responsible to community complaints and earning the community's respect as a provider of a positive environmental service.

An Environment Improvement Plan (EIP), developed in consultation with the community, is a way of providing a structured approach to addressing environmental matters. Operators who have an environment management system, EIP and environmental audit programme are eligible for accredited licensee status with EPA.

Risk Management

Operators, through training and total quality management procedures, should be encouraged to:

- Identify potential problems.
- Adopt a regular inspection and maintenance routine.
- Take appropriate corrective measures when problems do arise.
- Adopt operating and reporting procedures that seek to prevent problems happening again. Accidents, equipment failure, climatic events and human errors often result in incidents where people and the environment are at risk.

Emergency Preparedness

Dairy production plants should develop and maintain contingency plans. The plans should provide for the avoidance and control of spills, leakage or breakdowns so as to prevent pollution of the environment.

Contingency Planning

The preparation of such plans should include:

- Emergency holding procedures.
- Clean-up procedures.
- Action to minimize any adverse effects.
- Methods for disposal of spilled materials.
- Training of personnel in adequate identification of materials and correct operating procedures to avoid or minimize the likelihood of spills.

Other likely incidents which should be anticipated may include.

- Disruption of power supplies.
- Human error.
- Disruption caused by acts of nature—such as storms, flooding, fire.
- Plant breakdowns—including drain blockages, pump failure.
- Waste overloading the treatment process.
- Temporary or permanent loss of trained personnel.

Monitoring and reporting should be undertaken for performance purposes, and to ensure that plant and equipment are performing efficiently. Monitoring and reporting is an integral part of an operation's environment management plan and should be developed in consultation with the relevant authorities on a site-specific basis.

Records should be maintained of monitoring data and procedures should be reviewed periodically. Plant managers should carefully observe the environmental performance of their plants and should institute remedial action should problems arise. Monitoring of waste and other environmental indicators may include:

- Continuous monitoring of simple liquid, solid and gaseous waste indicators in processing streams or effluent.
- Key waste stream indicator monitoring on weekly to monthly basis or as specified in licence.
- Surface water and groundwater quality monitored monthly to quarterly and should include annual biological and bacteriological assessments where necessary, to check for any impacts on aquatic systems.

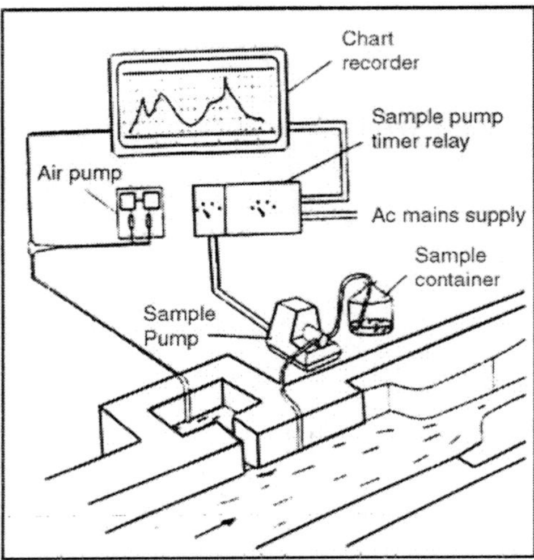

Fig. Continuous On-line Process Monitoring

- Soils in waste utilization areas should also be assessed for changes in chemical and physical characteristics, and to detect the onset of any soil degradation every five years.
- Key climatic parametres—such as rainfall and evapouration where irrigation is performed—should also be monitored on an annual basis.

Chapter 3

Hygiene in Food Processing

A high standard of hygiene is a prerequisite for safe food production, and the foundation on which HACCP and other safety management systems depend. Edited and written by some of the world's leading experts in the field, and drawing on the work of the prestigious European Hygienic Engineering and Design Group (EHEDG), Hygiene in food processing provides an authoritative and comprehensive review of good hygiene practice for the food industry.

- An authoritative and comprehensive review of good hygiene practice for the food industry
- Draws on the work of the prestigious European Hygienic Engineering and Design Group (EHEDG)
- Written and edited by world renowned experts in the field

Part 1: looks at the regulatory context, with chapters on the international context, regulation in the EU and the USA.

Part 2: looks at the key issue of hygienic design. After an introductory chapter on sources of contamination, there are chapters on plant design and control of airborne contamination. These are followed by a sequence of chapters on hygienic equipment design, including construction materials, piping systems, designing for cleaning in place and methods for verifying and certifying hygienic design.

Part 3: then reviews good hygiene practices, including cleaning and disinfection, personal hygiene and the management of foreign bodies and insect pests.

Drawing on a wealth of international experience and expertise, Hygiene in food processing will be a standard work for the food industry in ensuring safe food production.

CONTROL OF FOOD HAZARDS

Food business operators should control food hazards through the use of systems such as HACCP. They should:

- Identify any steps in their operations which are critical to the safety of food;
- Implement effective control procedures at those steps;
- Monitor control procedures to ensure their continuing effectiveness; and
- Review control procedures periodically, and whenever the operations change.

These systems should be applied throughout the food chain to control food hygiene throughout the shelf-life of the product through proper product and process design.

Control procedures may be simple, such as checking stock rotation calibrating equipment, or correctly loading refrigerated display units. In some cases a system based on expert advice, and involving documentation, may be appropriate.

KEY ASPECTS OF HYGIENE CONTROL SYSTEMS

Time and Temperature Control

Inadequate food temperature control is one of the most common causes of foodborne illness or food spoilage. Such controls include time and temperature of cooking, cooling, processing and storage. Systems should be in place to ensure that temperature is controlled effectively where it is critical to the safety and suitability of food.

Temperature control systems should take into account:

- The nature of the food, *e.g.* its water activity, pH, and likely initial level and types of micro-organisms;
- The intended shelf-life of the product;
- The method of packaging and processing; and
- How the product is intended to be used, *e.g.* further cooking/ processing or ready-to-eat.

Such systems should also specify tolerable limits for time and temperature variations. Temperature recording devices should be checked at regular intervals and tested for accuracy.

Specific Process Steps

Other steps which contribute to food hygiene may include, for example

- Chilling- thermal processing- irradiation
- Drying.
- Chemical preservation.
- Vacuum or modified atmospheric packaging

Microbiological and other Specifications

An effective way of ensuring the safety and suitability of food. Where microbiological, chemical or physical specifications are used in any food control system, such specifications should be based on sound scientific principles and state, where appropriate, monitoring procedures, analytical methods and action limits.

Microbiological Cross-Contamination

Pathogens can be transferred from one food to another, either by direct contact or by food handlers, contact surfaces or the air. Raw, unprocessed food should be effectively separated, either physically or by time, from ready-to-eat foods, with effective intermediate cleaning and where appropriate disinfection. Access to processing areas may need to be restricted or controlled. Where risks are particularly high, access to processing areas should be only via a changing facility. Personnel may need to be required to put on clean protective clothing including footwear and wash their hands before entering. Surfaces, utensils, equipment, fixtures and fittings should be thoroughly cleaned and where necessary disinfected after raw food, particularly meat and poultry, has been handled or processed.

Physical and Chemical Contamination

Systems should be in place to prevent contamination of foods by foreign bodies such as glass or metal shards from machinery, dust, harmful fumes and unwanted chemicals. In manufacturing and processing, suitable detection or screening devices should be used where necessary.

INCOMING MATERIAL REQUIREMENTS

No raw material or ingredient should be accepted by an establishment if it is known to contain parasites, undesirable microorganisms, pesticides, veterinary drugs or toxic, decomposed or extraneous substances which would not be reduced to an acceptable level by normal sorting and/or processing. Where appropriate, specifications for raw materials should be identified and applied.

Raw materials or ingredients should, where appropriate, be inspected and sorted before processing. Where necessary, laboratory tests should be made to establish fitness for use. Only sound, suitable raw materials or ingredients should be used.

Stocks of raw materials and ingredients should be subject to effective stock rotation.

PACKAGING

Packaging design and materials should provide adequate protection for products to minimize contamination, prevent damage, and accommodate proper labelling.

Packaging materials or gases where used must be non-toxic and not pose a threat to the safety and suitability of food under the specified conditions of storage and use. Where appropriate, reusable packaging should be suitably durable, easy to clean and, where necessary, disinfect.

WATER

In Contact with Food

Only potable water, should be used in food handling and processing, with the following exceptions:

- For steam production, fire control and other similar purposes not connected with food; and
- In certain food processes, *e.g.* chilling, and in food handling areas, provided this does not constitute a hazard to the safety and suitability of food (*e.g.* the use of clean sea water).

Water recirculated for reuse should be treated and maintained in such a condition that no risk to the safety and suitability of food results from its use. The treatment process should be effectively monitored. Recirculated water which has received no further treatment and water recovered from processing of food by evaporation or drying may be used, provided its use does not constitute a risk to the safety and suitability of food.

As an Ingredient

Potable water should be used wherever necessary to avoid food contamination.

Ice and Steam

Ice should be made from water that complies with section 4.4.1. Ice and steam should be produced, handled and stored to protect them from contamination.

Steam used in direct contact with food or food contact surfaces should not constitute a threat to the safety and suitability of food.

MANAGEMENT AND SUPERVISION

The type of control and supervision needed will depend on the size of the business, the nature of its activities and the types of food involved. Managers and supervisors should have enough knowledge of food hygiene principles and practices to be able to judge potential risks, take appropriate preventive and corrective action, and ensure that effective monitoring and supervision takes place.

DOCUMENTATION AND RECORDS

Where necessary, appropriate records of processing, production and distribution should be kept and retained for a period that exceeds the shelf-life of the product. Documentation can enhance the credibility and effectiveness of the food safety control system.

RECALL PROCEDURES

Managers should ensure effective procedures are in place to deal with any food safety hazard and to enable the complete, rapid recall of any implicated lot of the finished food from the market. Where a product has been withdrawn because of an immediate health hazard, other products which are produced under similar conditions, and which may present a similar hazard to public health, should be evaluated for safety and may need to be withdrawn. The need for public warnings should be considered.

Recalled products should be held under supervision until they are destroyed, used for purposes other than human consumption, determined to be safe for human consumption, or reprocessed in a manner to ensure their safety.

ESTABLISHMENT: MAINTENANCE AND SANITATION

Objective:

To establish effective systems to:

- *Ensure adequate and appropriate maintenance and cleaning;*
- *Control pests;*
- *Manage waste; and*
- *Monitor effectiveness of maintenance and sanitation procedures.*

Rationale:

To facilitate the continuing effective control of food hazards, pests, and other agents likely to contaminate food.

MAINTENANCE AND CLEANING

General

Establishments and equipment should be kept in an appropriate state of repair and condition to:

- Facilitate all sanitation procedures;
- Function as intended, particularly at critical steps;
- Prevent contamination of food, *e.g.* from metal shards, flaking plaster, debris and chemicals.

Cleaning should remove food residues and dirt which may be a source of contamination. The necessary cleaning methods and materials will depend on the nature of the food business. Disinfection may be necessary after cleaning. Cleaning chemicals should be handled and used carefully and in accordance with manufacturers' instructions and stored, where necessary, separated from food, in clearly identified containers to avoid the risk of contaminating food.

Cleaning Procedures and Methods

Cleaning can be carried out by the separate or the combined use of physical methods, such as heat, scrubbing, turbulent flow, vacuum cleaning or other methods that avoid the use of water, and chemical methods using detergents, alkalis or acids.

Cleaning procedures will involve, where appropriate:

- Removing gross debris from surfaces;

- Applying a detergent solution to loosen soil and bacterial film and hold them in solution or suspension;
- Rinsing with water which complies with section 4, to remove loosened soil and residues of detergent;
- Dry cleaning or other appropriate methods for removing and collecting residues and debris; and
- Where necessary, disinfection with subsequent rinsing unless the manufacturers' instructions indicate on a scientific basis that rinsing is not required.

CLEANING PROGRAMMES

Cleaning and disinfection programmes should ensure that all parts of the establishment are appropriately clean, and should include the cleaning of cleaning equipment.

Cleaning and disinfection programmes should be continually and effectively monitored for their suitability and effectiveness and where necessary, documented. Where written cleaning programmes are used, they should specify:

- Areas, items of equipment and utensils to be cleaned;
- Rresponsibility for particular tasks;
- Method and frequency of cleaning; and monitoring arrangements.

Where appropriate, programmes should be drawn up in consultation with relevant specialist expert advisors.

PEST CONTROL SYSTEMS

General

Pests pose a major threat to the safety and suitability of food. Pest infestations can occur where there are breeding sites and a supply of food. Good hygiene practices should be employed to avoid creating an environment conducive to pests. Good sanitation, inspection of incoming materials and good monitoring can minimize the likelihood of infestation and thereby limit the need for pesticides.

Preventing Access

Buildings should be kept in good repair and condition to prevent pest access and to eliminate potential breeding sites. Holes, drains and other places where pests are likely to gain access should be kept sealed. Wire mesh screens, for example on open windows, doors and ventilators, will reduce the problem of pest entry. Animals should, wherever possible, be excluded from the grounds of factories and food processing plants.

Harbourage and Infestation

The availability of food and water encourages pest harbourage and infestation. Potential food sources should be stored in pest-proof containers and/or stacked above the ground and away from walls. Areas both inside and outside food premises should be kept clean. Where appropriate, refuse should be stored in covered, pest-proof containers.

Monitoring and Detection

Establishments and surrounding areas should be regularly examined for evidence of infestation.

Eradication

Pest infestations should be dealt with immediately and without adversely affecting food safety or suitability. Treatment with chemical, physical or biological agents should be carried out without posing a threat to the safety or suitability of food.

Waste Management

Suitable provision must be made for the removal and storage of waste. Waste must not be allowed to accumulate in food handling, food storage, and other working areas and the adjoining environment except so far as is unavoidable for the proper functioning of the business.

Waste stores must be kept appropriately clean.

MONITORING EFFECTIVENESS

Sanitation systems should be monitored for effectiveness, periodically verified by means such as audit pre-operational inspections or, where appropriate, microbiological sampling of environment and food contact surfaces and regularly reviewed and adapted to reflect changed circumstances.

ESTABLISHMENT: PERSONAL HYGIENE

Objectives:

To ensure that those who come directly or indirectly into contact with food are not likely to contaminate food by:

- *Maintaining an appropriate degree of personal cleanliness;*
- *Behaving and operating in an appropriate manner.*

Rationale:

People who do not maintain an appropriate degree of personal cleanliness, who have certain illnesses or conditions or who behave inappropriately, can contaminate food and transmit illness to consumers.

HEALTH STATUS

People known, or suspected, to be suffering from, or to be a carrier of a disease or illness likely to be transmitted through food, should not be allowed to enter any food handling area if there is a likelihood of their contaminating food. Any person so affected should immediately report illness or symptoms of illness to the management.

Medical examination of a food handler should be carried out if clinically or epidemiologically indicated.

ILLNESS AND INJURIES

Conditions which should be reported to management so that any need for medical examination and/or possible exclusion from food handling can be considered, include:

- Jaundice.
- Diarrhoea.
- Vomiting.
- Fever sore throat with fever.
- Visibly infected skin lesions (boils, cuts, etc.).
- Discharges from the ear, eye or nose.

PERSONAL CLEANLINESS

Food handlers should maintain a high degree of personal cleanliness and, where appropriate, wear suitable protective clothing, head covering, and footwear. Cuts and wounds, where personnel are permitted to continue working, should be covered by suitable waterproof dressings. Personnel should always wash their hands when personal cleanliness may affect food safety, for example:

- At the start of food handling activities;
- Immediately after using the toilet; and
- After handling raw food or any contaminated material, where this could result in contamination of other food items; they should avoid handling ready-to-eat food, where appropriate.

PERSONAL BEHAVIOUR

People engaged in food handling activities should refrain from behaviour which could result in contamination of food, for example:

- Smoking;
- Spitting;
- Chewing or eating;
- Sneezing or coughing over unprotected food.

Personal effects such as jewellery, watches, pins or other items should not be worn or brought into food handling areas if they pose a threat to the safety and suitability of food.

VISITORS

Visitors to food manufacturing, processing or handling areas should, where appropriate, wear protective clothing and adhere to the other personal hygiene provisions in this section.

CHICKENS SHOWS PATHOGEN

While critics from the US Agriculture Department (USDA) have labelled the report "junk science" for not doing a wide enough sampling, the publishing of the survey by an influential consumer organisation is a direct attack on the food safety practices at leading poultry processors.

As well, the USDA's own figures show that samples in broilers, ground chicken and ground turkey testing positive for salmonella at US slaughter and processing plants have surged since 2002, indicating that companies might soon be pushed to do more to bring down the levels.

Broilers had the highest rates of salmonella, with 16.3 per cent of samples testing positive in 2005, up from 11.5 per cent in 2002,

USDA figures released earlier this year showed. The highest level was reached in 1998, when salmonella was found in 20 per cent of the broilers sampled.

Meanwhile the Consumer Reports survey tested 525 chickens — including samples from Perdue, Pilgrim's Pride and Tyson Foods. Campylobacter was present in 81 per cent of the chickens, salmonella in 15 per cent and both bacteria in 13 per cent. Only 17 per cent had neither pathogen.

That's the lowest percentage of clean birds in all four of the organisation's tests since 1998, and far less than the 51 per cent of clean birds it found in a 2003 survey, the organisation stated.

> "Leading chicken producers have stabilized the incidence of salmonella, but spiral-shaped campylobacter has wriggled onto more chickens than ever," Consumer Reports stated. "And although the US Department of Agriculture tests chickens for salmonella against a federal standard, it has not set a standard for campylobacter."

Overall, chickens labeled as organic or raised without antibiotics and costing $3 to $5 per pound were more likely to harbor salmonella than were conventionally produced broilers that cost more like $1 per pound, the study found.

Moreover, most of the bacteria we tested from all types of contaminated chicken showed resistance to one or more antibiotics, including some fed to chickens to speed their growth and those prescribed to humans to treat infections. Among all brands, 84 per cent of the salmonella and 67 per cent of the campylobacter organisms showed resistance to one or more antibiotics.

> "The findings suggest that some people who are sickened by chicken might need to try several antibiotics before finding one that works," the organisation concluded.

No major brand fared better than others overall. Foster Farms, Pilgrim's Pride, and Tyson chickens were lower in salmonella incidence than Perdue, but they were higher in campylobacter.

There was an exception to the poor showing of most premium chickens. As in its previous tests, Ranger—a no-antibiotics brand sold in the Northwest—was extremely clean. Of the 10 samples the organisation analyzed, none had salmonella, while two had campylobacter.

To keep contamination in check, processors follow procedures collectively known as Hazard Analysis and Critical Control Point (HACCP) analysis as a means of reducing the risks. HACCP requires companies to spell out where contamination could be controlled during processing, then build in procedures—such as scalding carcasses—to prevent it.

The report cited the example of Bell and Evans, which spent $30 million to modernize its processing plant in 2005. The sum included $9 million for a high-tech air-chill system designed in part to reduce cross-contamination.

The system whisks carcasses on two miles of track through chambers in which they're misted and chilled with air, then submerged in an antimicrobial dip. Tom Stone, the company's marketing director, says the measures helped reduce the rate of salmonella to less than 3 per cent in recent in-house tests of chickens done before packaging.

But in Consumer Reports tests of 28 store-bought chickens, five of the Bell and Evans samples had salmonella and 19 had campylobacter, the organisation stated. Other government studies back up the growth trend, while reporting significantly lower levels of the pathogens compared to the Consumer Reports survey.

SDA testing shows that Salmonella is found in about 12 per cent of broiler chickens in a sample of 3,242 nationwide in the third quarter of 2006. The industry association also cited a large-scale study by Dr. Norman Stern, an expert with the USDA Agricultural Research Service, which showed that 26 per cent of the chickens sampled had detectable level of Campylobacter.

Stern's study took place over 13 months, in 13 poultry complexes, and included 4,200 samples. This study was published in the Journal of Food Protection earlier this year.

> "As for the suggestion that bacteria that may be found on raw chicken are becoming more resistant to antibiotics, the fact is that Campylobacter is notoriously resistant to some antibiotics regardless of whether they are used in chicken or not," the associtation stated. "We are not aware of any actual human cases in which antibiotics have failed to work because of any usage in live chickens. The use of antibiotics in live chickens has declined sharply in recent years."

Another criticism of the Consumer Reports study came this week, with Reuters news service quoting a spokesman with the USDA as saying the Consumer Reports study was riddled with flaws such as a small sample size and uncertainty over the report's methodology. Steven Cohen, spokesman for the agency, said the report did not go back to all the stores used.

> "There is virtually nothing or any conclusion that anyone could draw from 500 samples," Reuters quotes Cohen as stating. "They're passing along junk science and calling it an investigation."

The federal Centers for Disease Control and Prevention (CDC) this month published USDA figures showing that the presence of Salmonella enterica serotype Enteritidis in broiler chicken carcass rinses collected from 2000 through 2005 increased four-fold over the period. The proportion of establishments with Salmonella Enteritidis–positive rinses increased nearly three fold during the same period.

> "We present US Department of Agriculture (USDA) Food Safety and Inspection Service (FSIS) Salmonella testing program data collected from 2000 to 2005 that suggest a need for interventions to prevent the emergence of this Salmonella serotype in broiler chickens in the United States," the CDC stated in its study.

Campylobacter and salmonella from all food sources sicken about 3.4 million Americans and kill about 700 annually, according to the latest estimates from the CDC.

The CDC said that in 2004, poultry was involved in 24 percent of infection outbreaks in which a single product was identified, up from 20 percent in 1998. In 2004, about 53 percent of campylobacter samples and 18 percent of salmonella samples were resistant to at least one antibiotic.

In July 2006, Food & Water Watch, an environmental health organization, published the names of 106 chicken processing plants that failed federal salmonella standards in at least one test period between 1998 and 2005.

In August 2006, the USDA reported that the rate of positive salmonella tests in broilers had jumped to 16.3 per cent in 2005, up from 11.5 per cent in 2002.

The use of thin walled packaging for food

In a survey of the segment, Applied Market Information (AMI) says thin-walled packaging has become an increasingly well delineated sub-segment of the rigid plastics market. In total in Europe the segment accounts for about two million tonnes of thermoplastics.

The report indicates some of the trends in this plastic packaging segment, as the market responds to higher raw material costs, and changes in demand for particular types of food products.

Thin walled packaging is used mainly to display goods for sale, but the format may also incorporate features for extending shelf-life and other functions. Thin walled packaging is defined to include yoghurt pots, butter and margarine containers, meat and fruit trays, blister packaging and similar containers. It excludes CD, DVD and blow moulded packaging. The largest single application of thin-walled packaging is dairy products, which account for 22 per cent of the market. The fastest growing application is in the chilled ready meals segment, in which demand is growing at about six per cent. While some of the applications for thin-walled packaging are mature others are evolving rapidly such as meat packaging, which is in the process of shifting from being packed within store to being packed centrally.

The fruit and vegetables segment accounts for eight per cent of the demand, meat, fish and poultry another eight per cent, frozen foods eight per cent, delicatessen products for seven per cent, bakery for five per cent, yellow fats for five per cent, chilled meals for four per cent and ambient products two per cent. Disposable products account for another 17 per cent of the demand and non-food products 14 per cent. In terms of composition, PET and PP are gaining market share, while polystyrene (PS) and polyvinyl chloride (PVC) are losing market share. Recently, plant-based and compostable polymers have emerged that offer brand owners exciting new opportunities to align sustainable and natural food propositions — such as organic food — with the packaging itself.

HYGIENIC PRODUCTION OF FOOD SOURCES

The potential effects of primary production activities on the safety and suitability of food should be considered at all times. In particular, this includes identifying any specific points in such activities where a high probability of contamination may exist and taking specific measures to minimize that probability.

Producers should as far as practicable implement measures to:

- Control contamination from air, soil, water, feedstuffs, fertilizers (including natural fertilizers), pesticides, veterinary drugs or any other agent used in primary production;
- Control plant and animal health so that it does not pose a threat to human health through food consumption, or adversely affect the suitability of the product; and
- Protect food sources from faecal and other contamination.

In particular, care should be taken to manage wastes, and store harmful substances appropriately. On-farm programmes which achieve specific food safety goals are becoming an important part of primary production and should be encouraged.

HANDLING, STORAGE AND TRANSPORT

Procedures should be in place to:

- Sort food and food ingredients to segregate material which is evidently unfit for human consumption;
- Dispose of any rejected material in a hygienic manner; and
- Protect food and food ingredients from contamination by pests, or by chemical, physical or microbiological contaminants or other objectionable substances during handling, storage and transport.

Care should be taken to prevent, so far as reasonably practicable, deterioration and spoilage through appropriate measures which may include controlling temperature, humidity, and/or other controls.

CLEANING, MAINTENANCE AND PERSONNEL HYGIENE AT PRIMARY PRODUCTION

Appropriate facilities and procedures should be in place to ensure that:

- Any necessary cleaning and maintenance is carried out effectively; and.
- An appropriate degree of personal hygiene is maintained.

ESTABLISHMENT: DESIGN AND FACILITIES

Objectives:

Depending on the nature of the operations, and the risks

associated with them, premises, equipment and facilities should be located, designed and constructed to ensure that:

- *Contamination is minimized;*
- *Design and layout permit appropriate maintenance, cleaning and disinfections and minimize air-borne contamination;*
- *Surfaces and materials, in particular those in contact with food, are non-toxic in intended use and, where necessary, suitably durable, and easy to maintain and clean;*
- *Where appropriate, suitable facilities are available for temperature, humidity and other controls; and there is effective protection against pest access and harbourage.*

Rationale:

Attention to good hygienic design and construction, appropriate location, and the provision of adequate facilities, is necessary to enable hazards to be effectively controlled.

LOCATION

Establishments

Potential sources of contamination need to be considered when deciding where to locate food establishments, as well as the effectiveness of any reasonable measures that might be taken to protect food. Establishments should not be located anywhere where, after considering such protective measures, it is clear that there will remain a threat to food safety or suitability. In particular, establishments should normally be located away from:

- Environmentally polluted areas and industrial activities which pose a serious threat of contaminating food;
- Areas subject to flooding unless sufficient safeguards are provided;
- Areas prone to infestations of pests;
- Areas where wastes, either solid or liquid, cannot be removed effectively.

Equipment

Equipment should be located so that it: permits adequate maintenance and cleaning; functions in accordance with its intended use; and facilitates good hygiene practices, including monitoring.

PREMISES AND ROOMS

Design and Layout

Where appropriate, the internal design and layout of food establishments should permit good food hygiene practices, including protection against cross-contamination between and during operations by foodstuffs.

INTERNAL STRUCTURES AND FITTINGS

Structures within food establishments should be soundly built of durable materials and be easy to maintain, clean and where appropriate, able to be disinfected. In particular the following specific conditions should be satisfied where necessary to protect the safety and suitability of food:

- The surfaces of walls, partitions and floors should be made of impervious materials with no toxic effect in intended use;
- Walls and partitions should have a smooth surface up to a height appropriate to the operation;
- Floors should be constructed to allow adequate drainage and cleaning;
- Ceilings and overhead fixtures should be constructed and finished to minimize the build up of dirt and condensation, and the shedding of particles;
- Windows should be easy to clean, be constructed to minimize the build up of dirt and where necessary, be fitted with removable and cleanable insect-proof screens. Where necessary, windows should be fixed;
- Doors should have smooth, non-absorbent surfaces, and be easy to clean and, where necessary, disinfect;
- Working surfaces that come into direct contact with food should be in sound condition, durable and easy to clean, maintain and disinfect. They should be made of smooth, non-absorbent materials, and inert to the food, to detergents and disinfectants under normal operating conditions.

TEMPORARY/MOBILE PREMISES AND VENDING MACHINES

Premises and structures covered here include market stalls, mobile sales and street vending vehicles, temporary premises in which food is handled such as tents and marquees. Such premises and structures should be sited, designed and constructed to avoid, as far as reasonably practicable, contaminating food and harbouring pests. In applying these specific conditions and requirements, any food hygiene hazards associated with such facilities should be adequately controlled to ensure the safety and suitability of food.

EQUIPMENT

General

Equipment and containers (other than once-only use containers and packaging) coming into contact with food, should be designed and constructed to ensure that, where necessary, they can be adequately cleaned, disinfected and maintained to avoid the contamination of food. Equipment and containers should be made of materials with no toxic effect in intended use. Where necessary, equipment should be durable and movable or capable of being disassembled to allow for maintenance, cleaning, disinfection, monitoring and, for example, to facilitate inspection for pests.

Food Control and Monitoring Equipment

In addition to the general requirements equipment used to cook, heat treat, cool, store or freeze food should be designed to achieve the required food temperatures as rapidly as necessary in the interests of food safety and suitability, and maintain them effectively. Such equipment should also be designed to allow temperatures to be monitored and controlled. Where necessary, such equipment should have effective means of controlling and monitoring humidity, air-flow and any other characteristic likely to have a detrimental effect on the safety or suitability of food. These requirements are intended to ensure that:

- Harmful or undesirable micro-organisms or their toxins are eliminated or reduced to safe levels or their survival and growth are effectively controlled;
- Where appropriate, critical limits established in HACCP-based plans can be monitored; and
- Temperatures and other conditions necessary to food safety and suitability can be rapidly achieved and maintained.

Containers for Waste and Inedible Substances

Containers for waste, by-products and inedible or dangerous substances, should be specifically identifiable, suitably constructed and, where appropriate, made of impervious material. Containers used to hold dangerous substances should be identified and, where appropriate, be lockable to prevent malicious or accidental contamination of food.

FACILITIES

Water Supply

An adequate supply of potable water with appropriate facilities for its storage, distribution and temperature control, should be available whenever necessary to ensure the safety and suitability of food.

Potable water should be as specified in the latest edition of WHO Guidelines for Drinking Water Quality, or water of a higher standard. Non-potable water (for use in, for example, fire control, steam production, refrigeration and other similar purposes where it would not contaminate food), shall have a separate system. Non-potable water systems shall be identified and shall not connect with, or allow reflux into, potable water systems.

Drainage and Waste Disposal

Adequate drainage and waste disposal systems and facilities should be provided. They should be designed and constructed so that the risk of contaminating food or the potable water supply is avoided.

Cleaning

Adequate facilities, suitably designated, should be provided for cleaning food, utensils and equipment. Such facilities should have an adequate supply of hot and cold potable water where appropriate.

Personnel Hygiene Facilities and Toilets

Personnel hygiene facilities should be available to ensure that an appropriate degree of personal hygiene can be maintained and to avoid contaminating food. Where appropriate, facilities should include:

- Adequate means of hygienically washing and drying hands, including wash basins and a supply of hot and cold (or suitably temperature controlled) water;
- Lavatories of appropriate hygienic design; and
- Adequate changing facilities for personnel.

Such facilities should be suitably located and designated.

Temperature Control

Depending on the nature of the food operations undertaken, adequate facilities should be available for heating, cooling, cooking, refrigerating and freezing food, for storing refrigerated or frozen foods, monitoring food temperatures, and when necessary, controlling ambient temperatures to ensure the safety and suitability of food.

Air Quality and Ventilation

Adequate means of natural or mechanical ventilation should be provided, in particular to:

- Minimize air-borne contamination of food, for example, from aerosols and condensation droplets;
- Control ambient temperatures;
- Control odours which might affect the suitability of food; and
- Control humidity, where necessary, to ensure the safety and suitability of food.

Ventilation systems should be designed and constructed so that air does not flow from contaminated areas to clean areas and, where necessary, they can be adequately maintained and cleaned.

Lighting

Adequate natural or artificial lighting should be provided to enable the undertaking to operate in a hygienic manner. Where necessary, lighting should not be such that the resulting colour is misleading. The intensity should be adequate to the nature of the operation. Lighting fixtures should, where appropriate, be protected to ensure that food is not contaminated by breakages.

Storage

Where necessary, adequate facilities for the storage of food, ingredients and non-food chemicals (*e.g.* cleaning materials, lubricants, fuels) should be provided. Where appropriate, food storage facilities should be designed and constructed to:

- Permit adequate maintenance and cleaning;
- Avoid pest access and harbourage;
- Enable food to be effectively protected from contamination during storage; and
- Where necessary, provide an environment which minimizes the deterioration of food (*e.g.* by temperature and humidity control).

The type of storage facilities required will depend on the nature of the food. Where necessary, separate, secure storage facilities for cleaning materials and hazardous substances should be provided.

CONTROL OF OPERATION

Objective:

To produce food which is safe and suitable for human consumption by:

- *Formulating design requirements with respect to raw materials, composition, processing, distribution, and consumer use to be met in the manufacture and handling of specific food items; and*
- *Designing, implementing, monitoring and reviewing effective control systems.*

Rationale:

To reduce the risk of unsafe food by taking preventive measures to assure the safety and suitability of food at an appropriate stage in the operation by controlling food hazards.

FOOD PROCESSING THROUGH COOKING METHODS

STEAMING

Steaming is a method of cooking using steam. Steaming is a preferred cooking method for health conscious individuals because no cooking oil is needed, thus resulting in a lower fat content. Steaming also results in a more nutritious food than boiling because fewer nutrients are destroyed or leached away into the water (which is usually discarded). It is also easier to avoid burning food when steaming. Steaming works by first boiling water, causing it to evapourate into steam; the steam then carries heat to the food, thus cooking the food.

In western cooking, steaming is most often used to cook vegetables, and only rarely to cook meats. By contrast, vegetables are seldom steamed in Chinese cuisine; vegetables are mostly stir fried or blanched instead. In Chinese cooking, steaming is used to cook many meat dishes, for example, steamed whole fish, steamed pork spare ribs, steamed ground pork or beef patties, steamed chicken, steamed goose etc. Other than meat dishes, many Chinese rice and wheat foods are steamed too. Examples include buns, Chinese steamed cakes etc. Steamed meat dishes (except some dim sum) are less common in Chinese restaurants than in traditional home cooking because meats usually require longer cooking time to steam than to stir fry.

The Chinese chefs developed an efficient method of restaurant cooking: big bamboo steaming baskets, each 1 m (3') in diameter and 10 cm (4") tall, can be stacked up on top of a wok like a chimney. The bottom of each basket is a grid which allows the steam from the wok to rise all the way to the top of the stack. In the kitchen of some dim sum restaurants, a steaming stack can be 20 levels high. The bottom level is removed when done and the entire stack simply shifted downward. This technique ensures a constant supply of freshly steamed dim sum.

Steaming at home can also be done with a wok. A shelf is put on the bottom of the wok, and a small steam basket or a dish of food is put on the shelf. Water is then filled to just below the dish or basket. The water is kept boiling, and a lid is placed over. Most vegetable dishes can be cooked in approximately 5 minutes using this method; most meat dishes, however, take longer than 20 minutes. A common alternative is to put the dish to be steamed on top of rice which is being cooked. A pot of rice which takes about 30 minutes to cook will then be ready at the same time as the steamed food.

Specialized steamers are often available for purchase; however, although they are more convenient, they are not necessarily better. Rice is traditionally steamed in the Lowcountry around Charleston, South Carolina, and specialized rice steamers are a common household cooking vessel in that area, although rather obscure elsewhere. A related technique is enclosing food in a container or material that will release steam when heated, such as clay pot cooking. A kind of steaming can be done outdoors by wrapping meat, poultry, or fish in banana leaves and burying it in hot sand or ash. Another form of outdoor steam cooking is covering a large piece of meat, poultry or fish in wet clay and placing it in a fire.

Double Steaming

Double steaming also called double boiling is a Chinese cooking technique to prepare delicate food such as bird nests, shark fins etc. The food is covered with water and put in a covered ceramic jar. The jar is then steamed for several hours. This technique ensures there is no loss of liquid or moisture (its essences) from the food being cooked, hence it is often used with expensive ingredients like Chinese herbal medicines.

Cantonese calls double steaming dan. Note that the Cantonese usage of this Chinese character deviates from its original meaning which is simmer or stew in Mandarin. This technique is also common in the neighbouring province of Fujian. There is another dessert dish called double steamed frog ovaries in a coconut recommended for women. The Chinese medicinal ingredients (including hasma), spices, and rock sugar are placed inside a young coconut to soak in the original coconut juice. The filled coconut is then double steamed for several hours. The whole coconut is served whole at the table after dinner. The contents and the inside wall of the coconut are scooped out to be consumed.

Grill (cooking)

There are multiple varieties of grills, with most falling into one of two categories: gas-fueled and charcoal. There is a great debate over the merits of charcoal or gas for use as the cooking method between grillers. Gas-fueled grills typically use propane (LP) or natural gas (NG) as their fuel source, with gas-flame either cooking food directly or heating grilling elements which in turn radiate the heat necessary to cook food. Gas grills are available in sizes ranging from small, single steak grills up to large, industrial sized restaurant grills which are able to cook enough meat to feed a hundred or more people. Gas grills are designed for either LP or NG, although it is possible to convert a grill from one gas source to another.

Charcoal grills typically use charcoal briquets as their fuel source. The briquets, when burned, will transform into embers radiating the heat necessary to cook food. One may say with certainty that E.G. Kingsford was the prime force behind the American grilling tradition. Kingsford was a relative of Henry Ford who saw that Ford's Model T production lines were producing a large amount of wood scraps that were just being discarded. Kingsford pitched a simple idea to Ford: Set up a charcoal manufacturing facility next to the assembly line and sell the charcoal, with the Ford name, in Ford dealerships. Ford, knowing a good idea when he saw one, immediately implemented Kingsford's idea. After Kingsford's death, the company was renamed Kingsford Charcoal Co. in his honor. Today, Kingsford charcoal is the dominant brand used by charcoal grillers.

Another personality in the charcoal grilling camp is George Stephen. The stereotypical American charcoal grill is a hollow, metal hemisphere with three legs and a small metal disc to catch ash, with a lower grate to hold the charcoal and an upper grate to hold the food to be cooked. George Stephen created the hemispherical grill design, jokingly called "Sputnik" by Stephen's neighbours. Stephen, a welder, worked for Weber Brothers Metal Works, a metal fabrication shop primarily concerned with welding steel spheres together to make buoys.

Stephen was tired of wind blowing ash onto his food when he grilled. One day he had an epiphany: he took the lower half of a buoy, welded three steel legs onto it, and fabricated a shallower hemisphere for use as a lid. He took the results home and within weeks was selling the grills first to his neighbours, then to customers, and finally started the Weber-Stephen Products Company. Weber grills come in many sizes, again, in small 14 inch diameter grills up to a full size 24 inch diameter grill. Grilling is a pervasive tradition in the United States. There are many cook-offs for steak grilling and barbecue (midwestern and southern style) around the United States with serious cash prizes involved in most. Almost all competition grillers use charcoal, most often in large, custom designed brick or steel grills. They can range from a few 55 gallon oil drums sawed lengthwise on their sides to make a lid and grill base, to large, vehicle sized grills made of brick, weighing nearly a ton.

Commonly grilled food and cooking methods

- Steaks. Pre-heat on high. This way the grill is hot and will scorch the outside of the steak. By the time it is "done" the outside will be blackened (but not burned) and you'll still have some pink in the middle.
- Shrimp. Medium/low heat. Don't overcook.
- Asparagus. Marinate in oil and salt.

Carbonation and sugarcane processing

Carbonatation is a chemical reaction where calcium hydroxide reacts with carbon dioxide and forms insoluble calcium carbonate:

$Ca(OH)_2 + CO_2 \rightarrow CaCO_3 + H_2O$

Carbonatation is a slow process that occurs in concrete where lime (calcium hydroxide) in the cement reacts with carbon dioxide from the air and forms calcium carbonate. Since carbonatation causes a lower pH (acidic) this may lead to corrosion of the steel reinforcing rods and damage to the construction.

Sugar Refining

The carbonatation process is used in the production of sugar from sugar bee', It is the introduction of milk of lime (calcium hydroxide suspension) and carbon dioxide enriched gas into the "raw juice", the sugar rich liquid prepared from the diffusion stage of the process, to form calcium carbonate and to precipitate and remove impurities. The whole process takes place in "carbonatation tanks" and processing time varies from 20 minutes to an hour.

Carbonatation involves the following effects:

- The increase in alkalinity coagulates proteins in the juice.
- Calcium carbonate absorbs colourants
- Alkalinity destroys some monosaccharide sugars, mostly glucose and fructose

The target is a large particle that naturally settles rapidly to leave a clear juice. The juice at the end is approximately 15 °Bx and 90% sucrose. The pH of the thin juice produced is a balance between removing as much calcium from the solution and the expected pH drop across later processing. If the juice goes acidic in the crystallisation stages then sucrose rapidly breaks down to glucose and fructose; not only do glucose and fructose affect crystallisation but they are melassagenic taking equivalent amounts of sucrose on to the molasses stage.

The carbon dioxide bubbled through the mixture forms calcium carbonate. The non-sugar solids are incorporated into the calcium carbonate particles and removed by natural (or assisted) sedimentation in tanks. There are several systems of carbonatation, named from the companies that first developed them. They differ in how the lime is introduced, the temperature and durations of each stage, and the separation of the solids from the liquid.

- Dorr (also Dorr-Oliver) - a continuous process using two tanks with recycling ("1st carbonatation") to build up particle size for natural flocculation. The recycling ratio is about 7:1. The particles are separated under gravity in a thickening stage in a vessel called a clarifier. The clear juice is then gassed further in another tank ("2nd carbonatation") and filtered. The concentrated mud (underflow) is filtered and/or pressed to recover more liquid. The Dorr process is low in maintenance and man-power but susceptible to frost damaged beet. It is favoured in the UK and the USA.
- DDS (Det Danske Sukkerfabrik - "The Danish Sugarfactory") — multistage process involving pre-liming where the pH of the juice is gradually increased to start precipitation of proteins, followed by addition of further lime and gas. The particles are removed at each stage by filtration.
- RT (Raffinière Tirlemontoise) - another multistage process with a pre-liming stage. Particles also removed by filtration.

Both DDS and RT processes are favoured by European factories. The clear juice from carbonatation is generally known as "thin juice". it may undergo pH adjustment with soda ash and addition of sulphur ("sulphitation") prior to the next stage which is concentration by multiple effect evapouration.

Blanching

Blanching is a cooking term that describes a process of food preparation wherein the food substance, usually a vegetable or fruit, is plunged into boiling water, removed after a brief, timed interval and finally plunged into iced water or placed under cold running water (shocked) to halt the cooking process.

Uses of blanching

- Peeling: Blanching loosens the skin on some fruits or nuts, such as onions, tomatoes, plums, peaches, or almonds.
- Flavour: Blanching enhances the flavor of some vegetables, such as broccoli, by releasing bitter acids stored in the cellular structure of the food.
- Appearance: Blanching enhances the colour of some (particularly green) vegetables by releasing gases trapped in the cellular material that obscure the greenness of the chorophyll. Since blanching is done - and halted - quickly, the heat does not have time to break down chlorophyll as well.
- Shelf life: Blanching neutralises bacteria and enzymes present in foods, thus delaying spoilage. This is often done as a preparatory step for freezing vegetables.

Blanching can also describe deep frying in oil at a lower temperature as with the initial cooking of French fries/chips.

Chapter 4

Heat Transfer in Food Processing

The principles of heat transfer are widely used in food processing in many items of equipment. It seems appropriate to discuss these under the various applications that are commonly encountered in nearly every food factory.

HEAT CONDUCTANCES IN PARALLEL

Heat conductances in parallel have a sandwich construction at right angles to the direction of the heat transfer, but with heat conductances in parallel, the material surfaces are parallel to the direction of heat transfer and to each other. The heat is therefore passing through each material at the same time, instead of through one material and then the next. This is illustrated in Figure.

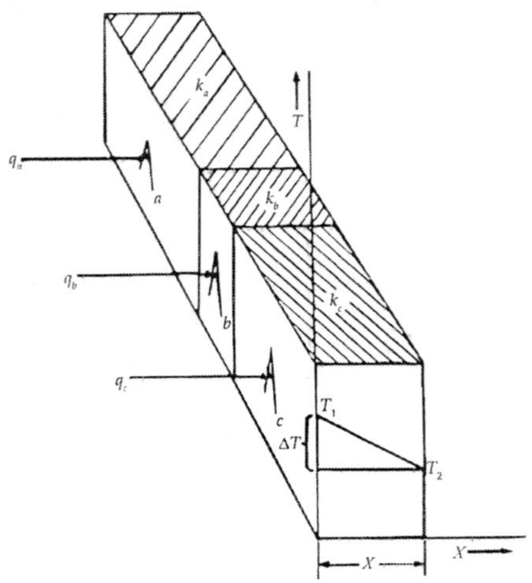

$$q_a = \frac{A_a \Delta T k_a}{X} \quad q_b = \frac{A_b \Delta T k_b}{X} \quad q_c = \frac{A_c \Delta T k_c}{X}$$

Fig. Heat Conductances in Parallel

An example is the insulated wall of a refrigerator or an oven, in which the walls are held together by bolts. The bolts are in parallel with the direction of the heat transfer through the wall: they carry most of the heat transferred and thus account for most of the losses.

Example: Heat Transfer in Walls of a Bakery Oven

The wall of a bakery oven is built of insulating brick 10 cm thick and thermal conductivity 0.22 J m^{-1} s^{-1} $°C^{-1}$. Steel reinforcing members penetrate the brick, and their total area of cross-section represents 1% of the inside wall area of the oven. If the thermal conductivity of the steel is 45 J m^{-1} s^{-1} $°C^{-1}$ calculate (a) the relative proportions of the total heat transferred through the wall by the brick and by the steel and (b) the heat loss for each m^2 of oven wall if the inner side of the wall is at 230°C and the outer side is at 25°C.

Applying equation $q = A\Delta Tk/x$, we know that ΔT is the same for the bricks and for the steel. Also x, the thickness, is the same.

(a) Consider the loss through an area of 1 m^2 of wall (0.99 m^2 of brick, and 0.01 m^2 of steel)

For brick $q_b = A_b \Delta T\, k_b/x$

$$= \frac{0.99(230-25)0.22}{0.10}$$

$= 446$ J s^{-1}

For steel $q_s = A_s \Delta T\, k_s/x$

$$= \frac{0.01(230-25)45}{0.10}$$

$= 923$ J s^{-1}

Therefore $q_b/q_s = 0.48$

(b) Total heat loss

$q = (q_b + q_s)$ per m^2 of wall

$= 446 + 923 = 1369$ J s^{-1}

Therefore percentage of heat carried by steel

$= (923/1369) \times 100$

$= 67\%$

HEAT CONDUCTANCES IN SERIES

Frequently in heat conduction, heat passes through several consecutive layers of different materials. For example, in a cold store wall, heat might pass through brick, plaster, wood and cork.

In this case, equation can be applied to each layer. This is illustrated in Figure.

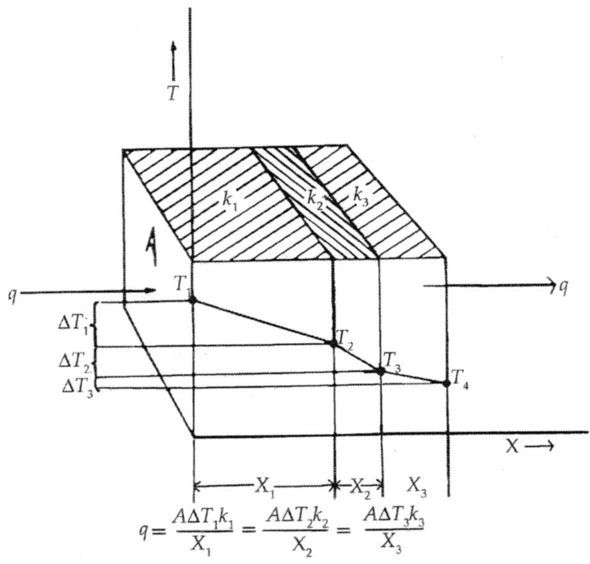

Fig. Heat conductances in Series

In the steady state, the same quantity of heat per unit time must pass through each layer.

$$q = A_1\Delta T_1 k_1/x_1 = A_2\Delta T_2 k_2/x_2$$
$$= A_3\Delta T_3 k_3/x_3 = \ldots\ldots\ldots$$

If the areas are the same,

$$A_1 = A_2 = A_3 = \ldots\ldots = A$$
$$q = A\Delta T_1 k_1/x_1 = A\Delta T_2 k_2/x_2$$
$$= A\Delta T_3 k_3/x_3 = \ldots\ldots\ldots$$

So $A\Delta T_1 = q(x_1/k_1)$

and

$$A\Delta T_2 = q(x_2/k_2)$$

and

$A\Delta T_3 \quad = q(x_3/k_3)$……

$A\Delta T_1 + A\Delta T_2 + A\Delta T_3 + \ldots = q(x_1/k_1) + q(x_2/k_2) + q(x_3/k_3) + \ldots$

$A(\Delta T_1 + \Delta T_2 + \Delta T_3 + ..) \quad = q(x_1/k_1 + x_2/k_2 + x_3/k_3 + \ldots)$

The sum of the temperature differences over each layer is equal to the difference in temperature of the two outside surfaces of the complete system, i.e.

$\Delta T_1 + \Delta T_2 + \Delta T_3 + \ldots \quad = \Delta T$

and since k_1/x_1 is equal to the conductance of the material in the first layer, C_1, and k_2/x_2 is equal to the conductance of the material in the second layer C_2,

$x_1/k_1 + x_2/k_2 + x_3/k_3 + \ldots \quad = 1/C_1 + 1/C_2 + 1/C_3 \ldots\ldots$

$= 1/U$

where U = the overall conductance for the combined layers, in J m^{-2} s^{-1} $°C^{-1}$

Therefore $\quad A\Delta T \; = q(1/U)$

And so $q \quad = UA\Delta T$

This is of the same form as equation but extended to cover the composite slab. U is called the overall heat-transfer coefficient, as it can also include combinations involving the other methods of heat transfer – convection and radiation.

Example: Heat Transfer in Cold Store Wall of Brick, Concrete and Cork

A cold store has a wall comprising 11 cm of brick on the outside, then 7.5 cm of concrete and then 10 cm of cork. The mean temperature within the store is maintained at –18°C and the mean temperature of the outside surface of the wall is 18°C.

Calculate the rate of heat transfer through the wall. The appropriate thermal conductivities are for brick, concrete and cork, respectively 0.69, 0.76 and 0.043 J m^{-1} s^{-1} $°C^{-1}$.

Determine also the temperature at the interfaces between the concrete and cork layers, and the brick and concrete layers.

For brick $\quad x_1/k_1 = 0.11/0.69 = 0.16.$

For concrete $\quad x_2/k_2 = 0.075/0.76 = 0.10.$

For cork $\quad x_3/k_3 = 0.10/0.043 = 2.33$

But $1/U \quad = x_1/k_1 + x_2/k_2 + x_3/k_3$

$= 0.16 + 0.10 + 2.33 = 2.59$

Therefore $U = 0.38 \text{ J m}^{-2} \text{ s}^{-1} {}^\circ\text{C}^{-1}$

$\Delta T = 18 - (-18) = 36^\circ\text{C}$,

$A = 1 \text{ m}^2$

$q = UA\Delta T$

$= 0.38 \times 1 \times 36$

$= 13.7 \text{ J s}^{-1}$

Further, $q = A_3 \Delta T_3 k_3 / x_3$

and for the cork wall $A_3 = 1 \text{ m}^2$, $x_3/k_3 = 2.33$

and

$q = 13.7 \text{ J s}^{-1}$

Therefore $13.7 = 1 \times \Delta T_3 \times 1/2.33$

from rearranging equation

$\Delta T_3 = 32^\circ\text{C}$.

But ΔT_3 is the difference between the temperature of the cork/concrete surface T_c and the temperature of the cork surface inside the cold store.

Therefore $T_c - (-18) = 32$

where T_c is the temperature at the cork/concrete surface

and so $T_c = 14^\circ\text{C}$.

If ΔT_1 is the difference between the temperature of the brick/concrete surface, T_b, and the temperature of the external air.

Then $13.7 = 1 \times \Delta T_1 \times 1/ 0.16 = 6.25\ \Delta T_1$

Therefore $18 - T_b = \Delta T_1 = 13.7/6.25 = 2.2$

so $T_b = 15.8\ ^\circ\text{C}$

Working it through shows approximate boundary temperatures: air/brick 18°C,brick/concrete 16°C, concrete/cork 14°C, cork/air –18°C.

This shows that almost all of the temperature difference occurs across the insulation (cork): and the actual intermediate temperatures can be significant especially if they lie below the temperature at which the atmospheric air condenses, or freezes.

HEAT TRANSFER FROM CONDENSING VAPOURS

The rate of heat transfer obtained when a vapour is condensing to a liquid is very often important. In particular, it occurs in the food industry in steam-heated vessels where the steam condenses and gives up its heat; and in distillation and evaporation where the vapours produced must be condensed. In condensation, the latent heat of vaporization is given up at constant temperature, the boiling temperature of the liquid.

Two generalized equations have been obtained:

(1) For condensation on vertical tubes or plane surfaces

$$h_v = 0.94[(k^3\rho^2 g/\mu) \times (\lambda/L\mu\Delta T)]^{0.25}$$

where λ(lambda) is the latent heat of the condensing liquid in J kg^{-1}, L is the height of the plate or tube and the other symbols have their usual meanings.

(2) For condensation on a horizontal tube

$$h_h = 0.72[(k^3\rho^2 g/\mu) \times (\lambda/D\Delta T)]^{0.25}$$

where D is the diameter of the tube.

These equations apply to condensation in which the condensed liquid forms a film on the condenser surface. This is called film condensation: it is the most usual form and is assumed to occur in the absence of evidence to the contrary. However, in some cases the condensation occurs in drops that remain on the surface and then fall off without spreading a condensate film over the whole surface. Since the condensate film itself offers heat transfer resistance, film condensation heat transfer rates would be expected to be lower than drop condensation heat transfer rates and this has been found to be true. Surface heat-transfer rates for drop condensation may be as much as ten times as high as the rates for film condensation.

The contamination of the condensing vapour by other vapours, which do not condense under the condenser conditions, can have a profound effect on overall coefficients. Examples of a non-condensing vapour are air in the vapours from an evaporator and in the jacket of a steam pan. The adverse effect of non-condensable vapours on overall heat transfer coefficients is due to the difference between the normal range of condensing heat transfer coefficients, 1200-12,000 J m^{-2} s^{-1} °C^{-1}, and the normal range of gas heat transfer coefficients with natural convection or low velocities, of about 6 J m^{-2} s^{-1} °C^{-1}.

Uncertainties make calculation of condensation coefficients difficult, and for many purposes it is near enough to assume the following coefficients:

For condensing steam 12,000 J m^{-2} s^{-1} °C^{-1}

For condensing ammonia 6,000 J m^{-2} s^{-1} °C^{-1}

For condensing organic liquids 1,200 J m^{-2} s^{-1} °C^{-1}

The heat-transfer coefficient for steam with 3% air falls to about 3500 J m^{-2} s^{-1} °C^{-1}, and with 6% air to about 1200 J m^{-2} s^{-1} °C^{-1}.

Example: Condensing Ammonia in a Refrigeration Plant

A steel tube of 1 mm wall thickness is being used to condense ammonia, using cooling water outside the pipe in a refrigeration plant. If the water side heat transfer coefficient is estimated at 1750 J m^{-2} s^{-1} °C^{-1} and the thermal conductivity of steel is 45 J m^{-1} s^{-1} °C^{-1}, calculate the overall heat-transfer coefficient.

Assuming the ammonia condensing coefficient, 6000 J m^{-2} s^{-1} °C^{-1}

$$1/U = 1/h_1 + x/k + 1/h_2$$
$$= 1/1750 + 0.001/45 + 1/6000$$
$$= 7.6 \times 10^{-4}$$
$$U = 1300 \text{ J m}^{-2} \text{ s}^{-1} \text{ °C}^{-1}.$$

HEAT REQUIREMENTS FOR VAPORIZATION

The energy, which must be supplied to vaporize the water at any temperature, depends upon this temperature. The quantity of energy required per kg of water is called the latent heat of vaporization, if it is from a liquid, or latent heat of sublimation if it is from a solid.

Example: Heat Energy in Air Drying

A food containing 80% water is to be dried at 100°C down to moisture content of 10%. If the initial temperature of the food is 21°C, calculate the quantity of heat energy required per unit weight of the original material, for drying under atmospheric pressure. The latent heat of vaporization of water at 100°C and at standard atmospheric pressure is 2257 kJ kg^{-1}. The specific heat capacity of the food is 3.8 kJ kg^{-1} °C^{-1} and of water is 4.186 kJ kg^{-1} °C^{-1}. Find also the energy requirement/kg water removed.

Calculating for 1 kg food

Initial moisture = 80%

800 g moisture are associated with 200 g dry matter.

Final moisture = 10 %,

100 g moisture are associated with 900 g dry .natter,

Therefore (100 × 200)/900 g = 22.2 g moisture are associated with 200 g dry matter.

1 kg of original matter must lose (800 – 22) g moisture = 778 g = 0.778 kg moisture.

Heat energy required for 1kg original material

= heat energy to raise temperature to 100°C + latent heat to remove water

= (100 – 21) × 3.8 + 0.778 × 2257

= 300.2 + 1755.9

= 2056 kJ.

Energy/kg water removed, as 2056 kJ are required to remove 0.778 kg of water,

= 2056/0.778

= 2643 kJ.

Steam is often used to supply heat to air or to surfaces used for drying. In condensing, steam gives up its latent heat of vaporization; in drying, the substance being dried must take up latent heat of vaporization to convert its liquid into vapour, so it might be reasoned that 1 kg of steam condensing will produce 1 kg vapour. This is not exactly true, as the steam and the food will in general be under different pressures with the food at the lower pressure. Latent heats of vaporization are slightly higher at lower pressures. In practice, there are also heat losses and sensible heat changes which may require to be considered.

Table. Latent Heat and Saturation Temperature of Water

Absolute Pressure (kPa)	Latent Heat of Vaporization (kJ kg^{-1})	Saturation Temperature (°C)
1	2485	7
2	2460	18
5	2424	33
10	2393	46
20	2358	60
20	2305	81
100	2258	99.6
101.35 (1 atm)	2257	100
110	2251	102
120	2244	105
200	2202	120
500	2109	152

Example: Heat Energy in Vacuum Drying

Using the same material as in Example 7.1, if vacuum drying is to be carried out at 60°C under the corresponding saturation pressure of 20 kPa abs. (or a vacuum of 81.4 kPa), calculate the heat energy required to remove the moisture per unit weight of raw material.

Heat energy required per kg raw material = heat energy to raise temperature to 60°C + latent heat of vaporization at 20 kPa abs.

$= (60 - 21) \times 3.8 + 0.778 \times 2358$

$= 148.2 + 1834.5$

$= 1983$ kJ.

In freeze drying the latent heat of sublimation must be supplied. Pressure has little effect on the latent heat of sublimation, which can be taken as 2838 kJ kg^{-1}.

Example: Heat Energy in Freeze Drying

If the foodstuff in the two previous examples were to be freeze dried at 0°C, how much energy would be required per kg of raw material, starting from frozen food at 0°C?

Heat energy required per kilogram of raw material= latent heat of sublimation

$= 0.778 \times 2838$

$= 2208$ kJ.

HEAT TRANSFER IN DRYING

We have been discussing the heat energy requirements for the drying process. The rates of drying are generally determined by the rates at which heat energy can be transferred to the water or to the ice in order to provide the latent heats, though under some circumstances the rate of mass transfer (removal of the water) can be limiting. All three of the mechanisms by which heat is transferred–conduction, radiation and convection–may enter into drying. The relative importance of the mechanisms varies from one drying process to another and very often one mode of heat transfer predominates to such an extent that it governs the overall process.

As an example, in air drying the rate of heat transfer is given by:

$q = h_s A(T_a - T_s)$

where q is the heat transfer rate in J s^{-1}, h_s is the surface heat-transfer coefficient J m^{-2} s^{-1} $°C^{-1}$, A is the area through which heat flow is taking place, m^2, T_a is the air temperature and T_s is the temperature of the surface which is drying, °C.

To take another example, in a roller dryer where moist material is spread over the surface of a heated drum, heat transfer occurs by conduction from the drum to the foodstuff, so that the equation is

$$q = UA(T_i - T_s)$$

where U is the overall heat-transfer coefficient, T_i is the drum temperature (usually very close to that of the steam), T_s is the surface temperature of the food (boiling point of water or slightly above) and A is the area of drying surface on the drum.

The value of U can be estimated from the conductivity of the drum material and of the layer of foodstuff. Values of U have been quoted as high as 1800 J m^{-2} s^{-1} $°C^{-1}$ under very good conditions and down to about 60 J m^{-2} s^{-1} $°C^{-1}$ under poor conditions.

In cases where substantial quantities of heat are transferred by radiation, it should be remembered that the surface temperature of the food may be higher than the air temperature. Estimates of surface temperature can be made using the relationships developed for radiant heat transfer although the actual effect of combined radiation and evaporative cooling is complex. Convection coefficients also can be estimated using the standard equations.

For freeze drying, energy must be transferred to the surface at which sublimation occurs. However, it must be supplied at such a rate as not to increase the temperature at the drying surface above the freezing point. In many applications of freeze drying, the heat transfer occurs mainly by conduction.

As drying proceeds, the character of the heat transfer situation changes. Dry material begins to occupy the surface layers and conduction must take place through these dry surface layers which are poor heat conductors so that heat is transferred to the drying region progressively more slowly.

DRYER EFFICIENCIES

Energy efficiency in drying is of obvious importance as energy consumption is such a large component of drying costs. Basically it is a simple ratio of the minimum energy needed to the energy actually consumed. But because of the complex relationships of the food, the water, and the drying medium which is often air, a number of efficiency measures can be worked out, each appropriate to circumstances and therefore selectable to bring out special features important in the particular process. Efficiency calculations are useful when assessing the performance of a dryer, looking for improvements, and in making comparisons between the various classes of dryers which may be alternatives for a particular drying operation.

Heat has to be supplied to separate the water from the food. The minimum quantity of heat that will remove the required water is that needed to supply the latent heat of evaporation, so one measure of efficiency is the ratio of that minimum to the energy actually provided for the process. Sensible heat can also be added to the minimum, as this added heat in the food often cannot be economically recovered.

Yet another useful measure for air drying such as in spray dryers, is to look at a heat balance over the air, treating the dryer as adiabatic with no exchange of heat with the surroundings. Then the useful heat transferred to the food for its drying corresponds to the drop in temperature in the drying air, and the heat which has to be supplied corresponds to the rise of temperature of the air in the air heater. So this adiabatic air-drying efficiency, h, can be defined by:

$$\eta = (T_1 - T_2)/(T_1 - T_a)$$

where T_1 is the inlet (high) air temperature into the dryer, T_2 is the outlet air temperature from the dryer, and T_a is the ambient air temperature. The numerator, the gap between T_1 and T_2, is a major factor in the efficiency.

Example: Efficiency of a Potato Dryer

A dryer reduces the moisture content of 100 kg of a potato product from 80% to 10% moisture. 250 kg of steam at 70 kPa gauge is used to heat 49,800 m^3 of air to 80°C, and the air is cooled to 71°C in passing through the dryer. Calculate the efficiency of the dryer. The specific heat of potato is 3.43 kJ kg^{-1} °C^{-1}. Assume potato enters at 24°C, which is also the ambient air temperature, and leaves at the same temperature as the exit air.

In 100 kg of raw material there is 80% moisture, that is 80 kg water and 20 kg dry material,

Total weight of dry product = 20 × (10/9)

= 22.2 kg

weight of water = (22.2 – 20)
= 2.2 kg.
water removed= (80 – 2.2)
= 77.8 kg.

Heat supplied to potato product = sensible heat to raise potato product temperature from 24°C to 71°C + latent heat of vaporization.

Now, the latent heat of vaporization corresponding to a saturation temperature of 71°C is 2331 kJ kg^{-1}

Heat (minimum) supplied/100 kg potato

$= 100 \times (71 - 24) \times 3.43 + 77.8 \times 2331$
$= 16 \times 10^3 + 181 \times 10^3$
$= 1.97 \times 10^5$ kJ.

Heat to evaporate water only $= 77.8 \times 2331$
$= 1.81 \times 10^5$ kJ

The specific heat of air is 1.0 J kg^{-1} $°C^{-1}$ and the density of air is 1.06 kg m^{-3} (Appendix 3)

Heat given up by air/100 kg potato

$= 1.0 \times (80 - 71) \times 49{,}800 \times 1.06$
$= 4.75 \times 10^5$ kJ.

The latent heat of steam at 70 kPa gauge is 2283 kJ kg^{-1}

Heat in steam $= 250 \times 2283$
$= 5.71 \times 10^5$ kJ.

Therefore (a) efficiency based on latent heat of vaporisation only:

$= (1.81 \times 10^5)/ (5.71 \times 10^5)$
$= 32\%$

(b) Efficiency assuming sensible heat remaining in food after drying is unavailable

$= (1.97 \times 10^5)/ (5.71 \times 10^5)$
$= 36\%$

(c) Efficiency based heat input and output, in drying air

$= (80 - 71)/ (80 - 24)$
$= 16\%$

Whichever of these is chosen depends on the objective for considering efficiency. For example in a spray dryer, the efficiency calculated on the air temperatures shows clearly and emphatically the advantages gained by operating at the highest feasible air inlet temperature and the lowest air outlet temperatures that can be employed in the dryer.

Examples of overall thermal efficiencies are:

Drum dryers 35-80%

Spray dryers 20-50%

Radiant dryers 30-40%

After sufficient energy has been provided to vaporize or to sublime moisture from the food, some way must be found to remove this moisture. In freeze-drying and vacuum systems it is normally convenient to condense the water to a liquid or a solid and then the vacuum pumps have to handle only the non-condensible gases. In atmospheric drying a current of air is normally used.

CONTINUOUS-FLOW HEAT EXCHANGERS

It is very often convenient to use heat exchangers in which one or both of the materials that are exchanging heat are fluids, flowing continuously through the equipment and acquiring or giving up heat in passing. One of the fluids is usually passed through pipes or tubes, and the other fluid stream is passed round or across these. At any point in the equipment, the local temperature differences and the heat transfer coefficients control the rate of heat exchange.

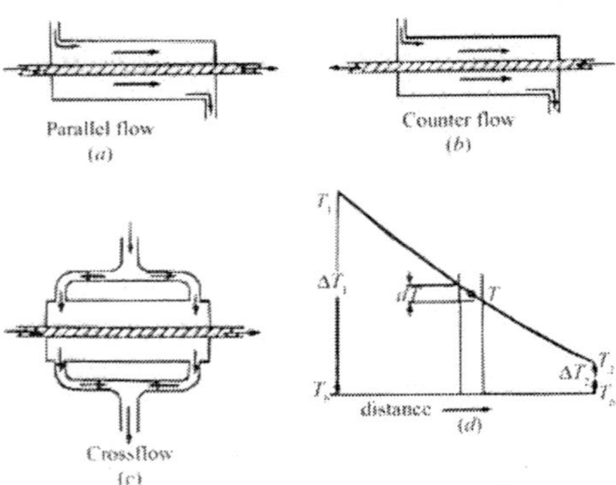

Fig. Heat Exchangers

The fluids can flow in the same direction through the equipment, this is called parallel flow; they can flow in opposite directions, called counter flow; they can flow at right angles to each other, called cross flow. Various combinations of these directions of flow can occur in different parts of the exchanger. Most actual heat exchangers of this type have a mixed flow pattern, but it is often possible to treat them from the point of view of the predominant flow pattern. Examples of these exchangers are illustrated in Figure.

In parallel flow, at the entry to the heat exchanger, there is the maximum temperature difference between the coldest and the hottest stream, but at the exit the two streams can only approach each other's temperature. In a counter flow exchanger, leaving streams can approach the temperatures of the entering stream of the other component and so counter flow exchangers are often preferred.

Applying the basic overall heat-transfer equation for the the heat exchanger heat transfer:

$q \quad = UA\,\Delta T$

uncertainty at once arises as to the value to be chosen for ΔT, even knowing the temperatures in the entering and leaving streams.

Consider a heat exchanger in which one fluid is effectively at a constant temperature, T_b as illustrated in Fig. 6.1(d). Constant temperature in one component can result either from a very high flow rate of this component compared with the other component, or from the component being a vapour such as steam or ammonia condensing at a high rate, or from a boiling liquid. The heat-transfer coefficients are assumed to be independent of temperature.

The rate of mass flow of the fluid that is changing temperature is G kg s^{-1}, its specific heat is c_p J kg^{-1} °C^{-1}. Over a small length of path of area dA, the mean temperature of the fluid is T and the temperature drop is dT. The constant temperature fluid has a temperature T_b. The overall heat transfer coefficient is U J m^{-2} s^{-1} °C^{-1}.

Therefore the heat balance over the short length is:

$c_pGdT \;= U(T - T_b)dA$

Therefore $\quad (U/)c_pG)\; dA = dT/(T - T_b)$

If this is integrated over the length of the tube in which the area changes from $A = 0$ to $A = A$, and T changes from T_1 to T_2, we have:

$U/(cpG)\,A \quad = \ln[(T_1 - T_b)/(T_2 - T_b)]$

(where ln = $\log_e$)

$= \ln\,(\Delta T_{1/}\,\Delta T_2)$

in which $\quad \Delta T_1 \;= (T_1 - T_b)$ and $\Delta T_2 = (T_2 - T_b)$

therefore $\quad c_pG \;= UA/\ln\,(\Delta T_{1/}\,\Delta T_2)$

From the overall equation, the total heat transferred per unit time is given by

$q \quad = UA\Delta T_m$

where ΔT_m is the mean temperature difference, but the total heat transferred per unit is also:

$q \quad = c_p G(T_1 - T_2)$

so $q \quad = UA\Delta T_m = c_p G(T_1 - T_2)$

$= UA/ \ln (\Delta T_{1/} \Delta T_2)] \times (T_1 - T_2)$

but $(T_1 - T_2)$ can be written $(T_1 - T_b) - (T_2 - T_b)$

so $(T_1 - T_2) = (\Delta T_1 - \Delta T_2)$

therefore $UA\Delta T_m = UA(\Delta T_1 - \Delta T_2)/ \ln (\Delta T_{1/} \Delta T_2)$

so that

$\Delta T_m = (\Delta T_1 - \Delta T_2)/ \ln (\Delta T_{1/} \Delta T_2)$

where ΔT_m is called the log mean temperature difference.

In other words, the rate of heat transfer can be calculated using the heat transfer coefficient, the total area, and the log mean temperature difference. This same result can be shown to hold for parallel flow and counter flow heat exchangers in which both fluids change their temperatures.

The analysis of cross-flow heat exchangers is not so simple, but for these also the use of the log mean temperature difference gives a good approximation to the actual conditions if one stream does not change very much in temperature.

Example: Cooling of Milk in a Pipe Heat Exchanger

Milk is flowing into a pipe cooler and passes through a tube of 2.5 cm internal diameter at a rate of 0.4 kg s^{-1}. Its initial temperature is 49°C and it is wished to cool it to 18°C using a stirred bath of constant 10°C water round the pipe. What length of pipe would be required? Assume an overall coefficient of heat transfer from the bath to the milk of 900 J m^{-2} s^{-1} $°C^{-1}$, and that the specific heat of milk is 3890 J kg^{-1} $°C^{-1}$.

Now

$q \quad = c_p G (T_1 - T_2)$

$= 3890 \times 0.4 \times (49 - 18)$

$= 48{,}240$ J s^{-1}

Also $q \quad = UA\Delta T_m$

$\Delta T_m = [(49 - 10) - (18 - 10)]/ \ln[(49 - 10)1(18 - 10)]$

$= 19.6°C.$

Therefore $\quad 48{,}240 \quad = 900 \times A \times 19.6$

$A \quad = 2.73 \text{ m}^2$

but $A \quad = \pi DL$

where L is the length of pipe of diameter D

Now$D \quad = 0.025$ m.

$L \quad = 2.73/(\pi \times 0.025)$

$\quad = 34.8$ m

This can be extended to the situation where there are two fluids flowing, one the cooled fluid and the other the heated fluid.

Working from the mass flow rates (kg s^{-1}) and the specific heats of the two fluids, the terminal temperatures can normally be calculated and these can then be used to determine ΔT_m and so, from the heat-transfer coefficients, the necessary heat-transfer surface.

Example: Water Chilling in a Counter Flow Heat Exchanger

In a counter flow heat exchanger, water is being chilled by a sodium chloride brine.

If the rate of flow of the brine is 1.8 kg s^{-1} and that of the water is 1.05 kg s^{-1}, estimate the temperature to which the water is cooled if the brine enters at –8°C and leaves at 10°C, and if the water enters the exchanger at 32°C.

If the area of the heat-transfer surface of this exchanger is 55 m^2, what is the overall heat-transfer coefficient? Take the specific heats to be 3.38 and 4.18 kJ kg^{-1} °C^{-1} for the brine and the water respectively.

With heat exchangers a small sketch is often helpful:

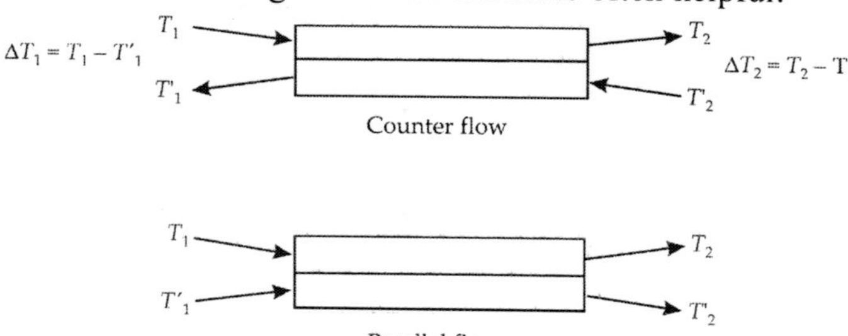

Fig. Diagrammatic heat exchanger

Figure shows three temperatures are known and the fourth T_{w2} (=T''_2 say on Fig) can be found from the heat balance:

By heat balance, heat loss in brine = heat gain in water

$$1.8 \times 3.38 \times [10 - (-8)] = 1.05 \times 4.18 \times (32 - T_{w2})$$

Therefore $T_{w2} = 7°C$.

And for counterflow

$\Delta T_1 = [32 - 10] = 22°C$

and

$\Delta T_2 = [7 - (-8)] = 15°C$.

Therefore $\Delta T_m = (22 - 15)/\ln(22/15)$

$= 7/0.382$

$= 18.3°C$.

For the heat exchanger

q = heat exchanged between fluids

= heat lost by brine = heat gain to water

= heat passed across heat transfer surface

$= UA\Delta T_m$

Therefore

$3.38 \times 1.8 \times 18 = U \times 55 \times 18.3$

$U = 0.11 \text{ kJ m}^{-2} \text{ °C}^{-1}$

$= 110 \text{ J m}^{-2} \text{ °C}^{-1}$

Parallel flow situations can be worked out similarly, making appropriate adjustments.

In some cases, heat-exchanger problems cannot be solved so easily; for example, if the heat transfer coefficients have to be calculated from the basic equations of heat transfer which depend on flow rates and temperatures of the fluids, and the temperatures themselves depend on the heat-transfer coefficients. The easiest way to proceed then is to make sensible estimates and to go through the calculations. If the final results are coherent, then the estimates were reasonable. If not, then make better estimates, on the basis of the results, and go through a new set of calculations; and if necessary repeat again until consistent results are obtained. For those with multiple heat exchangers to design, computer programmes are available.

JACKETED PANS

In a jacketed pan, the liquid to be heated is contained in a vessel, which may also be provided with an agitator to keep the liquid on the move across the heat-transfer surface.

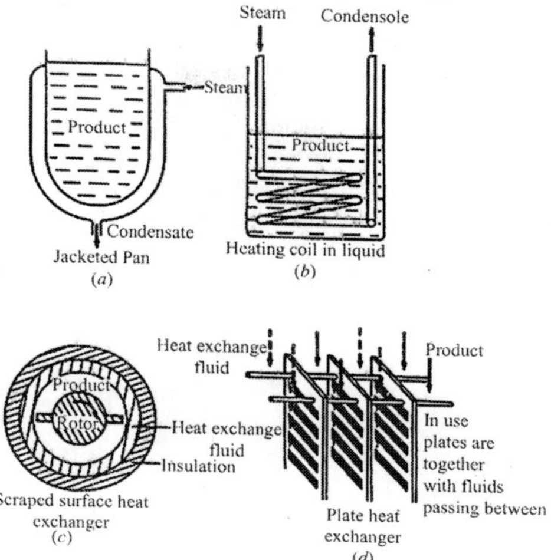

Fig. Heat Exchange Equipment

The source of heat is commonly steam condensing in the vessel jacket. Practical considerations of importance are:

- There is the minimum of air with the steam in the jacket.
- The steam is not superheated as part of the surface must then be used as a de-superheater over which low gas heat-transfer coefficients apply rather than high condensing coefficients.
- Steam trapping to remove condensate and air is adequate.

The action of the agitator and its ability to keep the fluid moved across the heat transfer surface are important. Some overall heat transfer coefficients are shown in Table.

Save for boiling water, which agitates itself, mechanical agitation is assumed. Where there is no agitation, coefficients may be halved.

Table. Some Overall Heat Transfer Coefficients in Jacketed Pans

Condensing fluid	Heated fluid	Pan material	Heat transfer Coefficients J m^{-2} s^{-1} $°C^{-1}$
Steam	Thin liquid	Cast-iron	1800
Steam	Thin liquid	Cast-iron	900
Steam	Paste	Stainless steel	300
Steam	Water, boiling	Copper	1800

Example: Steam required to Heat Pea Soup in Jacketed Pan

Estimate the steam requirement as you start to heat 50 kg of pea soup in a jacketed pan, if the initial temperature of the soup is 18°C and the steam used is at 100 kPa gauge. The pan has a heating surface of 1 m^2 and the overall heat transfer coefficient is assumed to be 300 J m^{-2} s^{-1} °C^{-1}.

From steam tables, saturation temperature of steam at 100 kPa gauge = 120°C and latent heat = l = 2202 kJ kg^{-1}.

$$q = UA\,\Delta T$$
$$= 300 \times 1 \times (120 - 18)$$
$$= 3.06 \times 10^4 \text{ J s}^{-1}$$

Therefore amount of steam

$$= q/\lambda = (3.06 \times 10^4)/(2.202 \times 10^6)$$
$$= 1.4 \times 10^{-2} \text{ kg s}^{-1}$$
$$= 1.4 \times 10^{-2} \times 3.6 \times 10^3$$
$$= 50 \text{ kg h}^{-1}.$$

This result applies only to the beginning of heating; as the temperature rises less steam will be consumed as ΔT decreases.

The overall heating process can be considered by using the analysis that led up to equation. A stirred vessel to which heat enters from a heating surface with a surface heat transfer coefficient which controls the heat flow, follows the same heating or cooling path as does a solid body of high internal heat conductivity with a defined surface heating area and surface heat transfer coefficient.

Example: Time to Heat Pea Soup in a Jacketed Pan

In the heating of the pan in Example, estimate the time needed to bring the stirred pea soup up to a temperature of 90°C, assuming the specific heat is 3.95 kJ kg^{-1} °C^{-1}.

From equation

$$(T_2 - T_a)/(T_1 - T_a) = \exp(-h_s At/ c\rho V)$$

T_a = 120°C (temperature of heating medium)

T_1 = 18°C (initial soup temperature)

T_2 = 90°C (soup temperature at end of time t)

h_s = 300 J m^{-2} s^{-1} °C^{-1}

A = 1 m^2, c = 3.95 kJ kg^{-1} °C^{-1}.

ρV = 50 kg

$$\text{Therefore } t = t = \frac{-3.95 \times 10^3 \times 50}{300 \times 1} \times \ln (90 - 120)/ (18 - 120)$$

$= (-658) \times (-1.22)$ s

$= 803$ s

$= 13.4$ min.

HEAT TRANSFER THEORY

Heat transfer is an operation that occurs repeatedly in the food industry. Whether it is called cooking, baking, drying, sterilizing or freezing, heat transfer is part of the processing of almost every food. An understanding of the principles that govern heat transfer is essential to an understanding of food processing.

Heat transfer is a dynamic process in which heat is transferred spontaneously from one body to another cooler body. The rate of heat transfer depends upon the differences in temperature between the bodies, the greater the difference in temperature, the greater the rate of heat transfer.

Temperature difference between the source of heat and the receiver of heat is therefore the driving force in heat transfer. An increase in the temperature difference, increases the driving force and therefore increases the rate of heat transfer. The heat passing from one body to another travels through some medium which in general offers resistance to the heat flow. Both these factors, the temperature difference and the resistance to heat flow, affect the rate of heat transfer. As with other rate processes, these factors are connected by the general equation:

Rate of transfer = Driving Force/ Resistance

For heat transfer:

Rate of heat transfer = Temperature difference/ heat flow resistance of medium

During processing, temperatures may change and therefore the rate of heat transfer will change. This is called unsteady state heat transfer, in contrast to steady state heat transfer when the temperatures do not change. An example of unsteady state heat transfer is the heating and cooling of cans in a retort to sterilize the contents. Unsteady state heat transfer is more complex since an additional variable, time, enters into the rate equations.

Heat can be transferred in three ways: by conduction, by radiation and by convection.

In conduction, the molecular energy is directly exchanged, from the hotter to the cooler regions, the molecules with greater energy communicating some of this energy to neighbouring molecules with less energy. An example of conduction is the heat transfer through the solid walls of a refrigerated store.

Radiation is the transfer of heat energy by electromagnetic waves, which transfer heat from one body to another, in the same way as electromagnetic light waves transfer light energy. An example of radiant heat transfer is when a foodstuff is passed below a bank of electric resistance heaters that are red-hot.

Convection is the transfer of heat by the movement of groups of molecules in a fluid. The groups of molecules may be moved by either density changes or by forced motion of the fluid. An example of convection heating is cooking in a jacketed pan: without a stirrer, density changes cause heat transfer by natural convection; with a stirrer, the convection is forced.

In general, heat is transferred in solids by conduction, in fluids by conduction and convection. Heat transfer by radiation occurs through open space, can often be neglected, and is most significant when temperature differences are substantial. In practice, the three types of heat transfer may occur together. For calculations it is often best to consider the mechanisms separately, and then to combine them where necessary.

HEAT CONDUCTION

In the case of heat conduction, the equation, rate = driving force/resistance, can be applied directly. The driving force is the temperature difference per unit length of heat-transfer path, also known as the temperature gradient. Instead of resistance to heat flow, its reciprocal called the conductance is used. This changes the form of the general equation to:

Rate of heat transfer = driving force × conductance,

that is:

$$dQ/dt = kA\ dT/dx$$

where dQ/dt is the rate of heat transfer, the quantity of heat energy transferred per unit of time, A is the area of cross-section of the heat flow path, dT/dx is the temperature gradient, that is the rate of change of temperature per unit length of path, and k is the thermal conductivity of the medium. Notice the distinction between thermal conductance, which relates to the actual thickness of a given material (k/x) and thermal conductivity, which relates only to unit thickness.

The units of k, the thermal conductivity, can be found from equation by transposing the terms

$$
\begin{aligned}
k &= dQ/dt \times 1/A \times 1/(dT/dx) \\
&= \text{J s}^{-1} \times \text{m}^{-2} \times 1/(°\text{C m}^{-1}) \\
&= \text{J m}^{-1}\ \text{s}^{-1}\ °\text{C}^{-1}
\end{aligned}
$$

Equation is known as the Fourier equation for heat conduction.

Note: Heat flows from a hotter to a colder body that is in the direction of the negative temperature gradient. Thus a minus sign should appear in the Fourier equation. However, in simple problems the direction of heat flow is obvious and the minus sign is considered to be confusing rather than helpful, so it has not been used.

THERMAL CONDUCTIVITY

On the basis of equation thermal conductivities of materials can be measured. Thermal conductivity does change slightly with temperature, but in many applications it can be regarded as a constant for a given material.

In general, metals have a high thermal conductivity, in the region 50-400 J m^{-1} s^{-1} °C^{-1}. Most foodstuffs contain a high proportion of water and as the thermal conductivity of water is about 0.7 J m^{-1} s^{-1}°C^{-1} above 0°C, thermal conductivities of foods are in the range 0.6 – 0.7 J m^{-1} s^{-1}°C^{-1}. Ice has a substantially higher thermal conductivity than water, about 2.3 J m^{-1} s^{-1}°C^{-1}. The thermal conductivity of frozen foods is, therefore, higher than foods at normal temperatures.

Most dense non-metallic materials have thermal conductivities of 0.5–2 J m^{-1} s^{-1}°C^{-1}. Insulating materials, such as those used in walls of cold stores, approximate closely to the conductivity of gases as they are made from non-metallic materials enclosing small bubbles of gas or air. The conductivity of air is 0.024 J m^{-1} s^{-1} °C^{-1} at 0°C, and insulating materials such as foamed plastics, cork and expanded rubber are in the range 0.03– 0.06 J m^{-1} s^{-1} °C^{-1}. Some of the new foamed plastic insulating materials have thermal conductivities as low as 0.026 J m^{-1} s^{-1} $°C^{-1}$. When using published tables of data, the units should be carefully checked. Mixed units, convenient for particular applications, are sometimes used and they may need to be converted.

CONDUCTION THROUGH A SLAB

If a slab of material, as shown in Figure, has two faces at different temperatures T_1 and T_2 heat will flow from the face at the higher temperature T_1 to the other face at the lower temperature T_2.

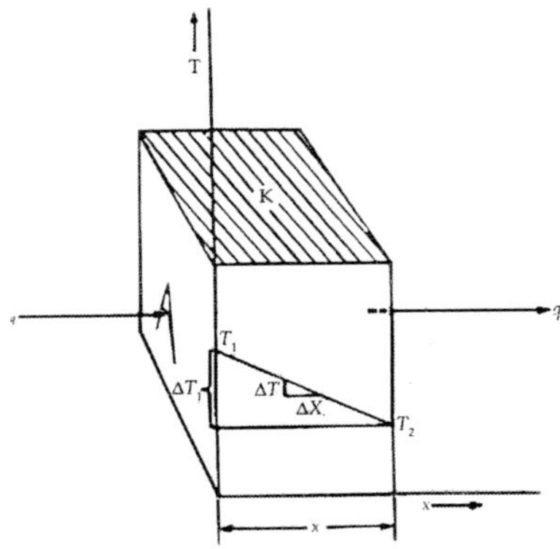

Fig. Heat Conduction Through a Slab

The rate of heat transfer is given by Fourier's equation:

$dQ/dt = kA\ \Delta T/\Delta x = kA\ dT/dx$

Under steady temperature conditions dQ/dt = constant, which may be called q:

and so $q = kA\ dT/dx$

but dT/dx, the rate of change of temperature per unit length of path, is given by $(T_1 - T_2)/x$ where × is the thickness of the slab,

so $q = kA(T_1 - T_2)/x$

or $q = kA\ \Delta T/x = (k/x)\ A\ \Delta T$

This may be regarded as the basic equation for simple heat conduction.

It can be used to calculate the rate of heat transfer through a uniform wall if the temperature difference across it and the thermal conductivity of the wall material are known.

Example: Rate of Heat Transfer in Cork

A cork slab 10 cm thick has one face at –12°C and the other face at 21°C. If the mean thermal conductivity of cork in this temperature range is 0.042 J m^{-1} s^{-1} °C^{-1}, what is the rate of heat transfer through 1 m^2 of wall?

T_1 = 21°C
T_2 = –12°C ΔT = 33°C
A = 1 m^2
k = 0.042 J m^{-1} s^{-1} °C^{-1}
x = 0.1 m

$$q = \frac{0.042}{0.1} \times 1 \times 33$$

= 13.9 J s^{-1}

SURFACE HEAT TRANSFER

Newton found, experimentally, that the rate of cooling of the surface of a solid, immersed in a colder fluid, was proportional to the difference between the temperature of the surface of the solid and the temperature of the cooling fluid. This is known as Newton's Law of Cooling, and it can be expressed by the equation, analogous to equation

$q = h_s A(T_a - T_s)$

where h_s is called the surface heat transfer coefficient, T_a is the temperature of the cooling fluid and T_s is the temperature at the surface of the solid. The surface heat transfer coefficient can be regarded as the conductance of a hypothetical surface film of the cooling medium of thickness x_f such that

$h_s = k_f x_f$

where k_f is the thermal conductivity of the cooling medium.

Following on this reasoning, it may be seen that h_s can be considered as arising from the presence of another layer, this time at the surface, added to the case of the composite slab considered previously. The heat passes through the surface, then through the various elements of a composite slab and then it may pass through a further surface film. We can at once write the important equation:

$q = A\Delta T[(1/h_{s1}) + x_1/k_1 + x_2/k_2 +.. + (1/h_{s2})]$
$= UA\Delta T$

where $1/U = (1/h_{s1}) + x_1/k_1 + x_2/k_2 +.. + (1/h_{s2})$

and h_{s1}, h_{s2} are the surface coefficients on either side of the composite slab, x_1, x_2... are the thicknesses of the layers making up the slab, and k_1, k_2.. are the conductivities of layers of thickness x_1.... The coefficient h_s is also known as the convection heat transfer coefficient and values for it will be discussed in detail under the heading of convection. It is useful at this point, however, to appreciate the magnitude of h_s under various common conditions and these are shown in Table.

Table. Approximate Range of Surface Heat Transfer Coefficients

	h (J m^{-2} s^{-1}°C^{-1})
Boiling liquids	2400–24,000
Condensing liquids	180–18,00
Still air	6
Moving air (3 m s^{-1})	30
Liquids flowing through pipes	1200–6000

Example: Heat Transfer in Jacketed Pan

Sugar solution is being heated in a jacketed pan made from stainless steel, 1.6 mm thick. Heat is supplied by condensing steam at 200 kPa gauge in the jacket. The surface transfer coefficients are, for condensing steam and for the sugar solution, 12,000 and 3000 J m^{-2} s^{-1} °C^{-1} respectively, and the thermal conductivity of stainless steel is 21 J m^{-1} s^{-1} °C^{-1}.

Calculate the quantity of steam being condensed per minute if the transfer surface is 1.4 m^2 and the temperature of the sugar solution is 83°C.

From steam tables, the saturation temperature of steam at 200 kPa gauge(300 kPa Absolute) = 134°C, and the latent heat = 2164 kJ kg^{-1}.

For stainless steel $x/k = 0.0016/21 = 7.6 \times 10^{-5}$

ΔT = (condensing temperature of steam) – (temperature of sugar solution)

= 134 – 83 = 51°C.

From equation.

$1/U = 1/12{,}000 + 7.6 \times 10^{-5} + 1/3000$

$U = 2032$ J m^{-2} s^{-1} °C^{-1}

and since $A = 1.4\ m^2$

$$q = UA\Delta T$$
$$= 2032 \times 1.4 \times 51$$
$$= 1.45 \times 10^5\ J\ s^{-1}$$

Therefore steam required

$$= 1.45 \times 10^5 / (2.164 \times 10^6)\ kg\ s^{-1}$$
$$= 0.067\ kg\ s^{-1}$$
$$= 4\ kg\ min^{-1}$$

UNSTEADY STATE HEAT TRANSFER

In food process engineering, heat transfer is very often in the unsteady state, in which temperatures are changing and materials are warming or cooling. Unfortunately, study of heat flow under these conditions is complicated. In fact, it is the subject for study in a substantial branch of applied mathematics, involving finding solutions for the Fourier equation written in terms of partial differentials in three dimensions. There are some cases that can be simplified and handled by elementary methods, and also charts have been prepared which can be used to obtain numerical solutions under some conditions of practical importance.

A simple case of unsteady state heat transfer arises from the heating or cooling of solid bodies made from good thermal conductors, for example a long cylinder, such as a meat sausage or a metal bar, being cooled in air. The rate at which heat is being transferred to the air from the surface of the cylinder is given by equation

$$q = dQ/dt = h_s A(T_s - T_a)$$

where T_a is the air temperature and T_s is the surface temperature.

Now, the heat being lost from the surface must be transferred to the surface from the interior of the cylinder by conduction. This heat transfer from the interior to the surface is difficult to determine but as an approximation, we can consider that all the heat is being transferred from the centre of the cylinder. In this instance, we evaluate the temperature drop required to produce the same rate of heat flow from the centre to the surface as passes from the surface to the air. This requires a greater temperature drop than the actual case in which much of the heat has in fact a shorter path. Assuming that all the heat flows from the centre of the cylinder to the outside, we can write the conduction equation

$dQ/dt = (k/L)A(T_c - T_s)$

where T_c is the temperature at the centre of the cylinder, k is the thermal conductivity of the material of the cylinder and L is the radius of the cylinder.

Equating these rates:

$h_s A(T_s - T_a) = (k/L)A(T_c - T_s)$

$h_s(T_s - T_a) = (k/L)(T_c - T_s)$

and so $h_s L/k = (T_c - T_s)/(T_s - T_a)$

To take a practical case of a copper cylinder of 15 cm radius cooling in air $k_c = 380$ J m^{-1} s^{-1} °C^{-1}, $h_s = 30$ J m^{-2} s^{-1}°C^{-1}, $L = 0.15$ m,

$(T_c - T_s)/(T_s - T_a) = (30 \times 0.15)/380$

$= 0.012$

In this case 99% of the temperature drop occurs between the air and the cylinder surface. By comparison with the temperature drop between the surface of the cylinder and the air, the temperature drop within the cylinder can be neglected. On the other hand, if the cylinder were made of a poorer conductor as in the case of the sausage, or if it were very large in diameter, or if the surface heat-transfer coefficient were very much larger, the internal temperature drops could not be neglected.

This simple analysis shows the importance of the ratio:

heat transfer coefficient at the surface $= h_s L/k$

heat conductance to the centre of the solid

This dimensionless ratio is called the Biot number (Bi) and it is important when considering unsteady state heat flow. When (Bi) is small, and for practical purposes this may be taken as any value less than about 0.2, the interior of the solid and its surface may be considered to be all at one uniform temperature. In the case in which (Bi) is less than 0.2, a simple analysis can be used, therefore, to predict the rate of cooling of a solid body.

Therefore for a cylinder of a good conductor, being cooled in air,

$dQ = h_s A(T_s - T_a)dt$

But this loss of heat cools the cylinder in accordance with the usual specific heat equation:

$dQ = c\rho V dT$

where c is the specific heat of the material of the cylinder, ρ is the density of this material and V is the volume of the cylinder.

Since the heat passing through the surface must equal the heat lost from the cylinder, these two expressions for dQ can be equated:

$$c\rho V dT = h_s A(T_s - T_a)dt$$

Integrating between $T_s = T_1$ and $T_s = T_2$, the initial and final temperatures of the cylinder during the cooling period, t, we have:

$$-h_s At/c\rho V = \log_e (T_2 - T_a)/(T_1 - T_a)$$

or $$(T_2 - T_a)/(T_1 - T_a) = \exp(-h_s At/c\rho V)$$

For this case, the temperatures for any desired interval can be calculated, if the surface transfer coefficient and the other physical factors are known. This gives a reasonable approximation so long as (Bi) is less than about 0.2. Where (Bi) is greater than 0.2, the centre of the solid will cool more slowly than this equation suggests. The equation is not restricted to cylinders, it applies to solids of any shape so long as the restriction in (Bi), calculated for the smallest half-dimension, is obeyed.

Charts have been prepared which give the temperature relationships for solids of simple shapes under more general conditions of unsteady-state conduction. These charts have been calculated from solutions of the conduction equation and they are plotted in terms of dimensionless groups so that their application is more general. The form of the solution is:

$$f\{(T - T_0)/(T_i - T_0)\} = F\{(kt/c\rho L^2)(h_s L/k)\}$$

where f and F indicate functions of the terms following, T_i is the initial temperature of the solid, T_0 is the temperature of the cooling or heating medium, T is the temperature of the solid at time t, $(kt/c\rho L^2)$ is called the Fourier number (Fo) (this includes the factor k/cr the thermal conductance divided by the volumetric heat capacity, which is called the thermal diffusivity) and $(h_s L/k)$ is the Biot number.

A mathematical outcome that is very useful in these calculations connects results for two- and three-dimensional situations with results from one-dimensional situations. This states that the two and three-dimensional values called $F(x, y)$ and $F(x, y, z)$ can be obtained from the individual one-dimensional results if these are $F(x)$, $F(y)$ and $F(z)$, by simple multiplication:

$$F(x, y) = F(x)F(y)$$

and

$$F(x, y, z) = F(x)\, F(y)\, F(z)$$

Using the above result, the solution for the cooling or heating of a brick is obtained from the product of three slab solutions. The solution for a cylinder of finite length, such as a can, is obtained from the product of the solution for an infinite cylinder, accounting for the sides of the can, and the solution for a slab, accounting for the ends of the can.

Example: Heat Transfer in Cooking Sausages

A process is under consideration in which large cylindrical meat sausages are to be processed in an autoclave. The sausage may be taken as thermally equivalent to a cylinder 30 cm long and 10 cm in diameter. If the sausages are initially at a temperature of 21°C and the temperature in the autoclave is maintained at 116°C, estimate the temperature of the sausage at its centre 2 h after it has been placed in the autoclave.

Assume that the thermal conductivity of the sausage is 0.48 J m^{-1} s^{-1} $°C^{-1}$, that its specific gravity is 1.07, and its specific heat is 3350 J kg^{-1} $°C^{-1}$. The surface heat-transfer coefficient in the autoclave to the surface of the sausage is 1200 J m^{-2} s^{-1} $°C^{-1}$.

This problem can be solved by combining the unsteady-state solutions for a cylinder with those for a slab, working from Figure.

(a) For the cylinder, of radius r = 5 cm (instead of L in this case)

$Bi = h_s r/k = (1200 \times 0.05)/0.48 = 125$

(Often in these systems the length dimension used as parameter in the charts is the half-thickness, or the radius, but this has to be checked on the graphs used.)

So $1/(Bi) = 8 \times 10^{-3}$

After 2 hours $t = 7200$ s

Therefore $Fo = kt/c\rho r^2 = (0.48 \times 7200)/[3350 \times 1.07 \times 1000 \times (0.05)^2]$

$= 0.39$

and so from Figure for the cylinder:

$(T - T_0)/(T_i - T_0) = 0.175$ = say, $F(x)$.

(b) For the slab the half-thickness 30/2 cm = 0.15 m

and so $Bi = h_s L/k = (1200 \times 0.15)/0.48 = 375$

$1/Bi = 2.7 \times 10^{-3}$

$t = 7200$ s

as before and

$kt/c\rho L^2 = (0.48 \times 7200)/[3350 \times 1.07 \times 1000 \times (0.15)^2]$

$= 4.3 \times 10^{-2}$

and so from Figure for the slab:

$(T - T_0)/(T_i - T_0) = 0.98 =$ say, $F(y)$

So overall $(T - T_0)/(T_i - T_0) = F(x)\,F(y)$

$= 0.175 \times 0.98$

$= 0.172$

Therefore $\dfrac{T_2 - 116}{21 - 116} = 0.172$

Therefore $T_2 = 100°C$

HEAT TRANSFER TO BOILING LIQUIDS

When the presence of a heated surface causes a liquid near it to boil, the intense agitation gives rise to high local coefficients of heat transfer. A considerable amount of experimental work has been carried out on this, but generalized correlations are still not very adequate.

It has been found that the apparent coefficient varies considerably with the temperature difference between the heating surface and the liquid. For temperature differences greater than about 20°C, values of h decrease, apparently because of blanketing of the heating surface by vapours. Over the range of temperature differences from 1°C to 20°C, values of h for boiling water increase from 1200 to about 60,000 J m^{-2} s^{-1} $°C^{-1}$. For boiling water under atmospheric pressure, the following equation is approximately true:

$$h = 50(\Delta T)^{2.5}$$

where ΔT is the difference between the surface temperature and the temperature of the boiling liquid and it lies between 2°C and 20°C.

In many applications the high boiling film coefficients are not of much consequence, as resistance in the heat source controls the overall coefficients.

CONVECTION HEAT TRANSFER

Convection heat transfer is the transfer of energy by the mass movement of groups of molecules. It is restricted to liquids and gases, as mass molecular movement does not occur at an appreciable speed in solids. It cannot be mathematically predicted as easily as can transfer by conduction or radiation and so its study is largely based on experimental results rather than on theory.

The most satisfactory convection heat transfer formulae are relationships between dimensionless groups of physical quantities. Furthermore, since the laws of molecular transport govern both heat flow and viscosity, convection heat transfer and fluid friction are closely related to each other.

Convection coefficients will be studied under two sections, firstly, natural convection in which movements occur due to density differences on heating or cooling; and secondly, forced convection, in which an external source of energy is applied to create movement. In many practical cases, both mechanisms occur together.

NATURAL CONVECTION

Heat transfer by natural convection occurs when a fluid is in contact with a surface hotter or colder than itself. As the fluid is heated or cooled it changes its density. This difference in density causes movement in the fluid that has been heated or cooled and causes the heat transfer to continue.

There are many examples of natural convection in the food industry. Convection is significant when hot surfaces, such as retorts which may be vertical or horizontal cylinders, are exposed with or without insulation to colder ambient air. It occurs when food is placed inside a chiller or freezer store in which circulation is not assisted by fans. Convection is important when material is placed in ovens without fans and afterwards when the cooked material is removed to cool in air.

It has been found that natural convection rates depend upon the physical constants of the fluid, density ρ, viscosity μ, thermal conductivity k, specific heat at constant pressure c_p and coefficient of thermal expansion β (beta) which for gases = $1/T$ by Charles' Law. Other factors that also affect convection-heat transfer are, some linear dimension of the system, diameter D or length L, a temperature difference term, ΔT, and the gravitational acceleration g since it is density differences acted upon by gravity that create circulation. Heat transfer rates are expressed in terms of a convection heat transfer coefficient h_c, which is part of the general surface coefficient h_s, in equation.

Experimentally, if has been shown that convection heat transfer can be described in terms of these factors grouped in dimensionless numbers which are known by the names of eminent workers in this field:

Nusselt number (Nu) $= (h_c D/k)$
Prandtl number (Pr) $= (c_p \mu/k)$
Grash of number (Gr) $= (D^3 \rho^2 g\ \beta \Delta\ T/\mu^2)$

and in some cases a length ratio (L/D).

If we assume that these ratios can be related by a simple power function we can then write the most general equation for natural convection:

$$(Nu) = K(Pr)^k (Gr)^m (L/D)^n$$

Experimental work has evaluated K, k, m, n, under various conditions. Once K, k, m, n are known for a particular case, together with the appropriate physical characteristics of the fluid, the Nusselt number can be calculated. From the Nusselt number we can find h_c and so determine the rate of convection-heat transfer by applying equation. In natural convection equations, the values of the physical constants of the fluid are taken at the mean temperature between the surface and the bulk fluid. The Nusselt and Biot numbers look similar: they differ in that for Nusselt, k and h both refer to the fluid, for Biot k is in the solid and h is in the fluid.

NATURAL CONVECTION EQUATIONS

These are related to a characteristic dimension of the body (food material for example) being considered, and typically this is a length for rectangular bodies and a diameter for spherical/cylindrical ones.

(1) Natural convection about vertical cylinders and planes, such as vertical retorts and oven walls

$$(\mathrm{Nu}) = 0.53\ (\mathrm{Pr.Gr})^{0.25} \text{ for } 10^4 < (\mathrm{Pr.Gr}) < 10^9$$
$$(\mathrm{Nu}) = 0.12(\mathrm{Pr.Gr})^{0.33} \text{ for } 10^9 < (\mathrm{Pr.Gr}) < 10^{12}$$

For air these equations can be approximated respectively by:

$$h_c = 1.3(\Delta T/L)^{0.25}$$
$$h_c = 1.8(\Delta \mathrm{D} T)^{0.25}$$

Equations are dimensional equations and are in standard units (ΔT in °C and L (or D) in metres and h_c in J m^{-2} s^{-1} °C^{-1}). The characteristic dimension to be used in the calculation of (Nu) and (Gr) in these equations is the height of the plane or cylinder.

(2) Natural convection about horizontal cylinders such as a steam pipe or sausages lying on a rack

(Nu) = $0.54(\text{Pr.Gr})^{0.25}$ for laminar flow in range $10^3 <$ (Pr.Gr) $< 10^9$.

Simplified equations can be employed in the case of air, which is so often encountered in contact with hotter or colder foods giving again:

For $10^4 <$ (Pr.Gr) $< 10^9$

$h_c = 1.3(DT/D)^{0.25}$

and for $10^9 <$ (Pr.Gr) $< 10^{12}$

$h_c = 1.8(\Delta T)^{0.33}$

(3) Natural convection from horizontal planes, such as slabs of cake cooling

The corresponding cylinder equations may be used, employing the length of the plane instead of the diameter of the cylinder whenever *D* occurs in (Nu) and (Gr). In the case of horizontal planes, cooled when facing upwards, or heated when facing downwards, which appear to be working against natural convection circulation, it has been found that half of the value of h_c in equations corresponds reasonably well with the experimental results.

Note carefully that the simplified equations are dimensional. Temperatures must be in °C and lengths in m and then h_c will be in J m^{-2} s^{-1} °C^{-1}. Values for σ, k and μ are measured at the film temperature, which is midway between the surface temperature and the temperature of the bulk liquid.

Example: Heat Loss from a Cooking Vessel

Calculate the rate of convection heat loss to ambient air from the side walls of a cooking vessel in the form of a vertical cylinder 0.9 m in diameter and 1.2 m high. The outside of the vessel insulation, facing ambient air, is found to be at 49°C and the air temperature is 17°C.

First it is necessary to establish the value of (Pr.Gr).

From the properties of air, at the mean film temperature, (49 + 17)/2, that is 33°C,

$\mu = 1.9 \times 10^{-5}$ N s m^{-2}, $c_p = 1.0$ kJ kg^{-1} °C^{-1}, $k = 0.025$ J m^{-1} s^{-1} °C^{-1}, $\beta = 1/308$, $\rho = 1.12$ kg m^{-3}.

From the conditions of the problem, characteristic dimension = height

$= 1.2$ m, $\Delta T = 32°C$.

Therefore (Pr.Gr) $= (c_p m\mu/k)(D^3\rho^2 g\ \beta\ \Delta T/\mu^2)$

$= (L^3\rho^2 g\ \beta\ \Delta T\ c_p)/(\mu k)$

$= [(1.2)^3 \times (1.12)^2 \times 9.81 \times 32 \times 1.0 \times 10^3)/(308 \times 1.9 \times 10^{-5} \times 0.025)$

$= 5 \times 10^9$

Therefore equation is applicable.

and so $h_c = 1.8\ \Delta T^{0.25} = 1.8(32)^{0.25}$

$= 4.3$ J m^{-2} s^{-1} °C^{-1}

Total area of vessel wall $= \pi DL = \pi \times 0.9 \times 1.2 = 3.4$ m^2

$\Delta T = 32°C$.

Therefore heat loss rate $= h_c A(T_1 - T_2)$

$= 4.3 \times 3.4 \times 32$

$= 468$ J s^{-1}

RADIATION HEAT TRANSFER

Radiation heat transfer is the transfer of heat energy by electromagnetic radiation. Radiation operates independently of the medium through which it occurs and depends upon the relative temperatures, geometric arrangements and surface structures of the materials that are emitting or absorbing heat.

The calculation of radiant heat transfer rates, in detail, is beyond the scope of this book and for most food processing operations a simplified treatment is sufficient to estimate radiant heat effects. Radiation can be significant with small temperature differences as, for example, in freeze drying and in cold stores, but it is generally more important where the temperature differences are greater. Under these circumstances, it is often the most significant mode of heat transfer, for example in bakers' ovens and in radiant dryers.

The basic formula for radiant-heat transfer is the Stefan-Boltzmann Law

$$q = A\sigma T^4$$

where T is the absolute temperature (measured from the absolute zero of temperature at –273°C, and indicated in Bold type) in degrees Kelvin (K) in the SI system, and σ (sigma) is the Stefan-Boltzmann constant = 5.73×10^{-8} J m^{-2} s^{-1}K^{-4} The absolute temperatures are calculated by the formula K = (°C + 273).

This law gives the radiation emitted by a perfect radiator (a black body as this is called though it could be a red-hot wire in actuality). A black body gives the maximum amount of emitted radiation possible at its particular temperature. Real surfaces at a temperature T do not emit as much energy as predicted by eqn. (5.8), but it has been found that many emit a constant fraction of it. For these real bodies, including foods and equipment surfaces, that emit a constant fraction of the radiation from a black body, the equation can be rewritten

$$q = \varepsilon A \sigma T^4$$

where ε (epsilon) is called the emissivity of the particular body and is a number between 0 and 1. Bodies obeying this equation are called grey bodies.

Emissivities vary with the temperature T and with the wavelength of the radiation emitted. For many purposes, it is sufficient to assume that for:

- Dull black surfaces (lamp-black or burnt toast, for example), emissivity is approximately 1;
- Surfaces such as paper/painted metal/wood and including most foods, emissivities are about 0.9;
- Rough unpolished metal surfaces, emissivities vary from 0.7 to 0.25;
- Polished metal surfaces, emissivities are about or below 0.05.

These values apply at the low and moderate temperatures which are those encountered in food processing.

Just as a black body emits radiation, it also absorbs it and according to the same law, equation.

Again grey bodies absorb a fraction of the quantity that a black body would absorb, corresponding this time to their absorptivity α (alpha). For grey bodies it can be shown that $\alpha = \varepsilon$. The fraction of the incident radiation that is not absorbed is reflected, and thus, there is a further term used, the reflectivity, which is equal to $(1 - \alpha)$.

RADIATION BETWEEN TWO BODIES

The radiant energy transferred between two surfaces depends upon their temperatures, the geometric arrangement, and their emissivities. For two parallel surfaces, facing each other and neglecting edge effects, each must intercept the total energy emitted by the other, either absorbing or reflecting it. In this case, the net heat transferred from the hotter to the cooler surface is given by:

$$q = AC\sigma (T_1^4 - T_2^4)$$

where $1/C = 1/\varepsilon_1 + 1/\varepsilon_2 - 1$, ε_1 is the emissivity of the surface at temperature T_1 and ε_2 is the emissivity of the surface at temperature T_2.

RADIATION TO A SMALL BODY FROM ITS SURROUNDINGS

In the case of a relatively small body in surroundings that are at a uniform temperature, the net heat exchange is given by the equation

$$q = A\varepsilon\sigma(T_1^4 - T_2^4)$$

where ε is the emissivity of the body, T_1 is the absolute temperature of the body and T_2 is the absolute temperature of the surroundings.

For many practical purposes in food process engineering, equation covers the situation; for example for a loaf in an oven receiving radiation from the walls around it, or a meat carcass radiating heat to the walls of a freezing chamber.

In order to be able to compare the various forms of heat transfer, it is necessary to see whether an equation can be written for radiant heat transfer similar to the general heat transfer equation. This means that for radiant heat transfer:

$$q = h_r A(T_1 - T_2) = h_r A\Delta T$$

where h_r is the radiation heat-transfer coefficient, T_1 is the temperature of the body and T_2 is the temperature of the surroundings. (The T would normally be the absolute temperature for the radiation, but the absolute temperature difference is equal to the Celsius temperature difference, because 273 is added and subtracted and so $(T_1 - T_2) = (T_1 - T_2) = \alpha T$.

Equating equations

$$q = h_r A(T_1 - T_2) = A\varepsilon\sigma(T_1^4 - T_2^4)$$

Therefore $h_r = \varepsilon\sigma(T_1^4 - T_2^4)/ (T_1 - T_2)$

$$= \varepsilon\sigma(T_1 + T_2)(T_1^2 + T_2^2)$$

If $T_m = (T_1 + T_2)/2$, we can write $T_1 + e = T_m$ and $T_2 - e = T_m$ where

$$2e = T_1 - T_2$$

also

$(T_1 + T_2) = 2\ T_m$

and then

$(T_1^2 + T_2^2) = T_m^2 - 2eT_m + e^2 + T_m^2 + 2eT_m + e^2$

$= 2T_m^2 + 2e^2$

$= 2T_m^2 + (T_1 - T_2)^2/2$

Therefore $h_r = \varepsilon\sigma(2T_m)[2T_m^2 + (T_1 - T_2)^2/2]$

Now, if $(T_1 - T_2) \leftrightarrow T_1$ or T_2, that is if the difference between the temperatures is small compared with the numerical values of the absolute temperatures, we can write:

$h_r \approx \varepsilon\sigma\ 4T_m^3$

and so

$q = h_r A\ \Delta T$

$= (\varepsilon \times 5.73 \times 10^{-8} \times 4 \times T_m^3) \times A\ \Delta T$

$= 0.23e\ (T_{m/}\ 100)^3 A\ \Delta T$

Example: Radiation Heat Transfer to Loaf of Bread in an Oven

Calculate the net heat transfer by radiation to a loaf of bread in an oven at a uniform temperature of 177°C, if the emissivity of the surface of the loaf is 0.85, using equation. Compare this result with that obtained by using equation. The total surface area and temperature of the loaf are respectively 0.0645 m^2 and 100°C.

$q = A\varepsilon\sigma(T_1^4 - T_2^4)$

$= 0.0645 \times 0.85 \times 5.73 \times 10^{-8}\ (450^4 - 373^4)$

$= 68.0\ J\ s^{-1}$.

By equation

$q = 0.23e\ (T_m/100)^3 A\ DT$

$= 0.23 \times 0.85\ (411/100)^3 \times 0.0645 \times 77$

$= 67.4\ J\ s^{-1}$.

Notice that even with quite a large temperature difference, equation gives a close approximation to the result obtained using equation.

Chapter 5

Process of Enzymes Activity

Enzymes are catalysts that optimize cell activity while minimizing the amount of energy needed to achieve a specific reaction. Enzymes are also energized protein molecules found in every living cell, and are necessary for life. There are over 2000 known enzymes each of which is involved with one specific chemical reaction.

They are any of various proteins, originating from living cells and capable of producing certain chemical changes in organic substances by catalytic action, such as digestion. These proteins, and their function (s), are determined by their shape. In cells and organisms, most reactions are catalyzed by enzymes, which are regenerated during the course of a reaction.

Biological catalysts are physiologically important because they speed up rates of reactions that would otherwise be too slow to support life. Our bodies naturally produce digestive and metabolic enzymes as they are needed. Specifically, the pancreas produces enzymes that break down foods into nutrients the body can use for energy and other bodily functions. The names of enzymes often include the substrate or substance on which they act, joined with an-ase ending.

For example, lactase acts upon lactose and maltase acts on maltose to produce glucose. Sometimes they are named for their reaction product, for example, sucrase is often called invertase, because invertase is the result of the reaction of sucrose. Enzymes can also bear a name the describes the reaction that is catalyzed. An example of this includes the use of the name oxidase, because oxidase is involved in an oxidation reaction.

ENZYME CLASSIFICATIONS

Traditionally, enzymes were simply assigned names by the investigator who discovered the enzyme. As knowledge expanded, systems of enzyme classification became more comprehensive and complex. Currently enzymes are grouped into six functional classes by the International Union of Biochemists (I.U.B.). These rules give each enzyme a unique number. The I.U.B. system also specifies a textual name for each enzyme. The enzyme's name is comprised of the names of the substrate(s), the product(s) and the enzyme's functional class. Because many enzymes, such as alcohol dehydrogenase, are widely known in the scientific community by their common names, the change to I.U.B.-approved nomenclature has been slow. In everyday usage, most enzymes are still called by their common name.

Number	Classification	Biochemical Properties
1	Oxidoreductases	Act on many chemical groupings to add or remove hydrogen atoms.
2	Transferases	Transfer functional groups between donor and acceptor molecules. Kinases are specialized transferases that regulate metabolism by transferring phosphate from ATP to other molecules.
3	Hydrolases	Add water across a bond, hydrolyzing it.
4.	Lyases	Add water, ammonia or carbon dioxide across double bonds, or remove these elements to produce double bonds.
5	Isomerases	Carry out many kinds of isomerization: L to D isomerizations, mutase reactions (shifts of chemical groups) and others.
6	Ligases	Catalyze reactions in which two chemical groups are joined (or ligated) with the use of energy from ATP.

Enzymes are also classified on the basis of their composition. Enzymes composed wholly of protein are known as simple enzymes in contrast to complex enzymes, which are composed of protein plus a relatively small organic molecule. Complex enzymes are also known as holoenzymes. In this terminology the protein component is known as the apoenzyme, while the non-protein component is known as the coenzyme or prosthetic group where prosthetic complex in which the small organic molecule is bound to the apoenzyme by covalent bonds; when the binding between the apoenzyme and non-protein components is non-covalent, the small organic molecule is called a coenzyme. Many prosthetic groups and coenzymes are water-soluble derivatives of vitamins. It should be noted that the main clinical symptoms of dietary vitamin insufficiency generally arise from the malfunction of enzymes, which lack sufficient cofactors derived from vitamins to maintain homeostasis. The non-protein component of an enzyme may be as simple as a metal ion or as complex as a small non-protein organic molecule. Enzymes that require a metal in their composition are known as metalloenzymes if they bind and retain their metal atom(s) under all conditions with very high affinity. Those which have a lower affinity for metal ion, but still require the metal ion for activity, are known as metal-activated enzymes.

ENZYMES AND LIFE PROCESSES

The living cell is the site of tremendous biochemical activity called metabolism. This is the process of chemical and physical change which goes on continually in the living organism. Build-up of new tissue, replacement of old tissue, conversion of food to energy, disposal of waste materials, reproduction - all the activities that we characterize as "life." This building up and tearing down takes place in the face of an apparent paradox. The greatest majority of these biochemical reactions do not take place spontaneously. The phenomenon of catalysis makes possible biochemical reactions necessary for all life processes. Catalysis is defined as the acceleration of a chemical reaction by some substance which itself undergoes no permanent chemical change. The catalysts of biochemical reactions are enzymes and are responsible for bringing about almost all of the chemical reactions in living organisms. Without enzymes, these reactions take place at a rate far too slow for the pace of metabolism. The oxidation of a fatty acid to carbon dioxide and water is not a gentle process in a test tube - extremes of pH, high temperatures and corrosive chemicals are required. Yet in the body, such a reaction takes place smoothly and rapidly within a narrow range of pH and temperature. In the laboratory, the average protein must be boiled for about 24 hours in a 20% HCl solution to achieve a complete breakdown. In the body, the breakdown takes place in four hours or less under conditions of mild physiological temperature and pH.

CHEMICAL NATURE OF ENZYMES

All known enzymes are proteins. They are high molecular weight compounds made up principally of chains of amino acids linked together by peptide bonds.

Peptide Bond

$$R-\underset{NH_2}{\overset{H}{\underset{|}{\overset{|}{C}}}}-\boxed{CO-NH}-\underset{COOH}{\overset{H}{\underset{|}{\overset{|}{C}}}}-R'$$

where:

$$R'-\underset{COOH}{\overset{H}{\underset{|}{\overset{|}{C}}}}-NH_2$$

and

$$R-\underset{COOH}{\overset{H}{\underset{|}{\overset{|}{C}}}}-NH_2$$

Represent two typical amino acids

Fig. Typical Protein Structure - Two Amino Acids Joined by a Peptide Bond.

Enzymes can be denatured and precipitated with salts, solvents and other reagents. They have molecular weights ranging from 10,000 to 2,000,000. Many enzymes require the presence of other compounds - cofactors - before their catalytic activity can be exerted. This entire active complex is referred to as the holoenzyme; *i.e.*, apoenzyme (protein portion) plus the cofactor (coenzyme, prosthetic group or metal-ion-activator) is called the holoenzyme.

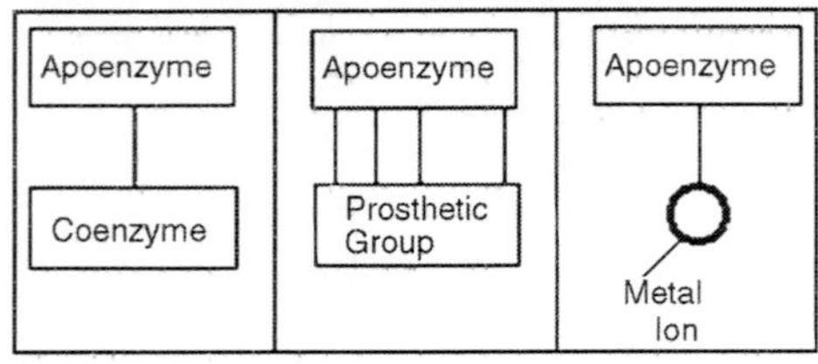

Fig. Holoenzymes- Apoenzymes Plus Various Types of Cofactors.

Apoenzyme + Cofactor = Holoenzyme

According to Holum, the cofactor may be:

- *A coenzyme*: A non-protein organic substance which is dialyzable, thermostable and loosely attached to the protein part.

- *A prosthetic group*: An organic substance which is dialyzable and thermostable which is firmly attached to the protein or apoenzyme portion.
- *A metal-ion-activator*: These include K^+, Fe^{++}, Fe^{+++}, Cu^{++}, Co^{++}, Zn^{++}, Mn^{++}, Mg^{++}, Ca^{++}, and Mo^{+++}.

NATURE OF ENZYMES

Enzymes are one of the most interesting and important substances found in nature. Firts, it's important to realise that enzymes are not living things. They are inanimate-like minerals. But unlike minerals, they are made by living cells. If we were to look inside a cell we would see many different activities going on. There would be some molecules joining together and others breaking apart. These activities keep the cell alive and enzymes make these activities possible. That's why every cell of every living creature on Earth produces enzymes. In a broad sense, there are two types of enzyme. Some that help join specific molecules together to form new molecules. Others that help break specific molecules apart into separate molecules.

Enzymes play many important roles ouside the cell as well. One of the best examples of this is the digestive system. For instance, it is enzymes in your digestive system that break food down in your digestive system break food down into small molecules that can be absorbed by the body. Some enzymes in your digestive system break down starch, some proteins and others break down fats.

FOUR THINGS TO REMEMBER ABOUT ENZYMES

- *Enzymes are specific*: An enzyme that is able to break fat down would not be able to dissolve protein or starch. Enzymes perform only one specific job. That means an enzyme can do its job with very few side effects. It also explains why there are so many different types of enzyme. To date, 3,000 different types have been identified and there are many more waiting to be discovered.
- *Enzymes are catalysts*: While it is true that an enzyme can only perform one speific job, it is important to know that one enzyme can perform that same job over and over again, millions of times, without being consumed in the process.

And enzymes do their job best in the mild ph and temperature conditions found in nature.

- *Enzymes are efficient*: Not only do enzymes work hard, they also work with blinding speed. For instance, there is an enzyme in the liver that helps hydrogern peroxide break down into water and oxygen. What's amazing is that one enzyme can process 5 million hydrogen peroxide molecules in one minute.
- *Enzymes are natural*: Enzymes are proteins. Like all other proteins, enzymes are organic. Once they have done their job, enzymes break down swiftly and can be absorbed back into nature.

ENZYMES IN INDUSTRY

Many people believe enzyme technology is fairly new. However, that's not the case. Enzymes have been used by man from the dawn of civilization. As long as people have been eating bread and cheese, and drinking wine and beer-they have been using enzymes. That's because enzymes help to make these products.

CHEESE

Take cheese for example. Most likely, cheese was discovered by ancient hunters who stored milk in the stomachs of slaughtered calves. When exposed to heat, the milk inside the container would turn into a solid-cheese. The reason for this is simple. Calves have an enzyme called chymosin in their stomachs. At ambient temperature chymosin will break down milk protein, causing milk to separate into curd and whey. If you strain away the watery whey you are left with a soft tangy curd or cheese. Chymosin derived from calf stomachs has been used in cheese making ever since.

Detergent

In more recent times, other commercial uses for enzymes have been found. Laundry detergents are a good example. As mentioned, there are certain enzymes which dissolve proteins and others which dissolve fat. Proteins and fat make up two of the major causes of stains on cloting. Grass, blood and egg are all protein stains. Lipstick, frying oil, butter, sauces, and tough stains on cuffs and collars are fat stains. to remove these stains without enzymes is difficult and requires a lot of washing at high temperatures with a lot of detergent.

Enzymes can remove these stains better, faster and in a way that is a lot kinder to the clothing and the environment. The enzymes in detergent dissolve stains in the same natural way they help digest food in your stomach-specifically and efficintly. Just a tiny amount of enzymes can dissolve stains that would require much larger amounts of detergent alone. Because of this they can reduce the amount of detergent actually found in laundry detergent. And since enzymes work best in mild conditions, washing machines using enzymatic detergants can be set at lower tempreatures, reducing electricity consumption by as much as a third. In the future, enzymes could replace more and more of the harsh chemicals found in detergents.

Many other Industries

These two examples represent just the tip of the iceberg. Enzymes are contributing to industry in many other areas. They are starting to replace petroleum-based solvents used to make vegetable seed oil; replacing harsh acids used in the production of glucose products such as corn syrup; taking over part of the role played by clorine in the paper industry; and replacing sulphides used in the tanning industry. Enzymes are also being used instead of pumice stones in the stonewashing of jeans.

Enzymes help Industry and Nature

Today enzymes are helping industry make the products used by society in a way that is less harmful to the environment. In the cases mentioned above, enzymatic processes are, in general, safer and more environment-friendly than the traditional processes they replace. That is why enzymes in particular and biotechnology in general hold such promise.

How Enzymes are Produced

Looking back to the cheese example above, chymosin is an enzyme used to turn milk into cheese. And chymosin is only found in the stomachs of young calves (as well as young sheep, goats, and a few other farm animals). Up until the 60s all the world's cheese was made from chymosin extracted from the stomacns of slaughtered calves. Then two things happened. The demand for cheese increased and the demand for alf meat decreased. Soon there were not enough slaughtered calf stomachs to produce enough high-quality chymosin required by the expanding cheese industry. To solve the chymosin shortage, it would have been very wasteful and very expensive to raise and slaughter calves just to extract a relatively small amount of chymosin from their stomachs. The cheese industry and the enzyme researchers began to look for another type of organism that produced chymosin.

Microorganisms are Natural Enzyme Factories

The researchers wanted to find an organism that was easier and less expensive to raise than a calf. An organism that reproduced quickly and wouldn't require a lot of space and food. So they began looking among the smallest and simplest organisms they knew: micoorganisms. Microorganisms are very small living organisms. Some examples are bacteria, fungi and yeast. They live in soil or water in every corner of the earth. Because they are so small, they are obviously a lot less complex than a calf. Even so, a simple microorganism can produce many different types of enzyme. In fact, it doesn't take a complex organism to create a complex enzyme-as was proved by a simple fungus named Mucor. This fungus produced a chymosin-like enzyme that was almost the same as that produced by calves. But finding the right enzyme is only half the battle.

Microorganisms often aren't Suited for "life on the Farm"

Once identified a microorganism needs to be "farmed" in a controlled setting. Most microorganisms can grow and multiply quite well in liquid. So large tanks are used called fermentation tanks to grow microorganisms. Ideally, a microorganism will grow fast and produce a lot of the desired enzyme at mild temperatures consuming inexpensive nutrients. But like most things in life, the ideal microorganism is hard to come by. As it is, most microorganisms found in the wild are not so well suited for domestication in fermentation tanks.

Some only produce tiny quantities of the desired enzyme, or they may produce undesirable by-products, or take a long time to grow, or require a growth environment that is very difficult and expensive to maintain. The result is that although we often find ourselves with a microorganism that can produce the enzyme needed the cost for cultivating it is prohibitive. That's where genetic engineering can help

Function and Structure

Enzymes are very efficient catalysts for biochemical reactions. They speed up reactions by providing an alternative reaction pathway of lower activation energy. Like all catalysts, enzymes take part in the reaction-that is how they provide an alternative reaction pathway. But they do not undergo permanent changes and so remain unchanged at the end of the reaction. They can only alter the rate of reaction, not the position of the equilibrium. Most chemical catalysts catalyse a wide range of reactions. They are not usually very selective. In contrast enzymes are usually highly selective, catalysing specific reactions only. This specificity is due to the shapes of the enzyme molecules. Many enzymes consist of a protein and a non-protein (called the cofactor). The proteins in enzymes are usually globular. The intra-and intermolecular bonds that hold proteins in their secondary and tertiary structures are disrupted by changes in temperature and pH. This affects shapes and so the catalytic activity of an enzyme is pH and temperature sensitive.

Cofactors may be:

- Organic groups that are permanently bound to the enzyme (prosthetic groups)
- Cations-positively charged metal ions (activators), which temporarily bind to the active site of the enzyme, giving an intense positive charge to the enzyme's protein
- Organic molecules, usually vitamins or made from vitamins (coenzymes), which are not permanently bound to the enzyme molecule, but combine with the enzyme-substrate complex temporarily.

HOW ENZYMES WORK

For two molecules to react they must collide with one another. They must collide in the right direction (orientation) and with sufficient energy. Sufficient energy means that between them they have enough energy to overcome the energy barrier to reaction. This is called the activation energy. Enzymes have an active site. This is part of the molecule that has just the right shape and functional groups to bind to one of the reacting molecules. The reacting molecule that binds to the enzyme is called the substrate. An enzyme-catalysed reaction takes a different 'route'. The enzyme and substrate form a reaction intermediate. Its formation has a lower activation energy than the reaction between reactants without a catalyst.

A simplified picture:

Route A reactant 1 + reactant 2 → product

Route Breactant 1 + enzyme → intermediate

intermediate + reactant 2 → product + enzyme

So the enzyme is used to form a reaction intermediate, but when this reacts with another reactant the enzyme reforms.

LOCK AND KEY HYPOTHESIS

This is the simplest model to represent how an enzyme works. The substrate simply fits into the active site to form a reaction intermediate.

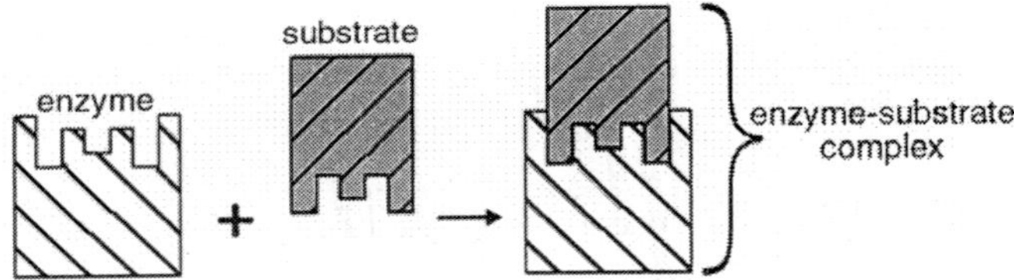

Induced Fit Hypothesis

In this model the enzyme molecule changes shape as the substrate molecules gets close. The change in shape is 'induced' by the approaching substrate molecule. This more sophisticated model relies on the fact that molecules are flexible because single covalent bonds are free to rotate.

Factors Affecting Catalytic Activity of Enzymes

Temperature

As the temperature rises, reacting molecules have more and more kinetic energy. This increases the chances of a successful collision and so the rate increases. There is a certain temperature at which an enzyme's catalytic activity is at its greatest. This optimal temperature is usually around human body temperature (37.5 °C) for the enzymes in human cells.

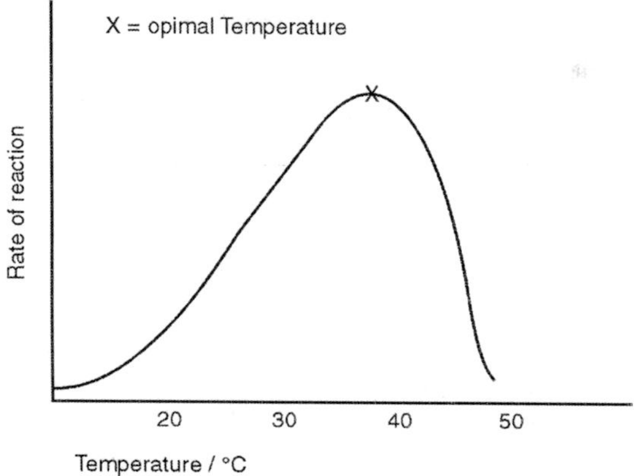

pH

Below this temperature the enzyme structure begins to break down (denature) since at higher temperatures intra-and intermolecular bonds are broken as the enzyme molecules gain even more kinetic energy. Each enzyme works within quite a small pH range. There is a pH at which its activity is greatest (the optimal pH). This is because changes in pH can make and break intra-and intermolecular bonds, changing the shape of the enzyme and, therefore, its effectiveness.

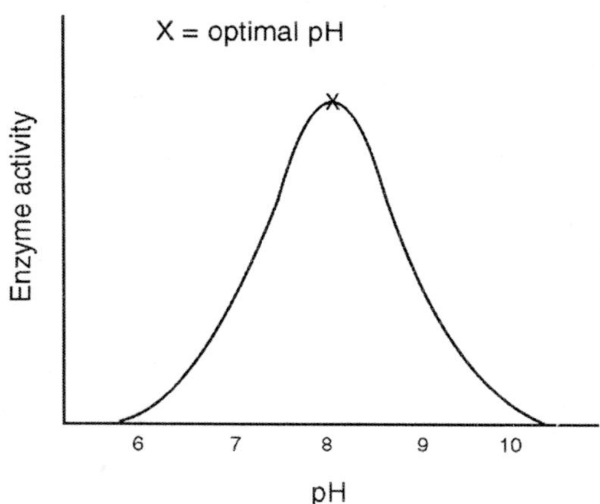

Concentration of Enzyme and Substrate

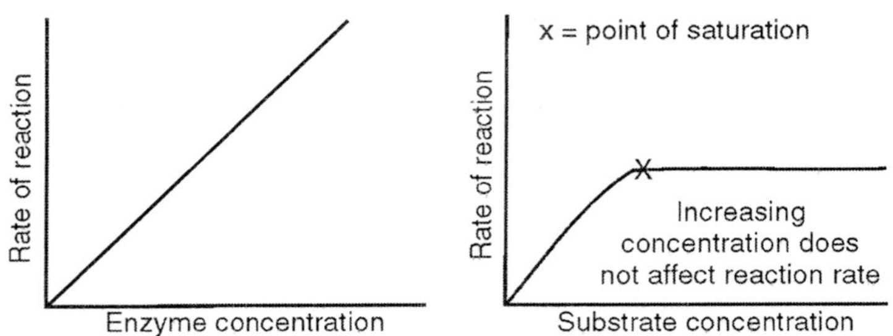

The rate of an enzyme-catalysed reaction depends on the concentrations of enzyme and substrate. As the concentration of either is increased the rate of reaction increases. For a given enzyme concentration, the rate of reaction increases with increasing substrate concentration up to a point, above which any further increase in substrate concentration produces no significant change in reaction rate.

This is because the active sites of the enzyme molecules at any given moment are virtually saturated with substrate. The enzyme/ substrate complex has to dissociate before the active sites are free to accommodate more substrate Provided that the substrate concentration is high and that temperature and pH are kept constant, the rate of reaction is proportional to the enzyme concentration.

Inhibition of Enzyme Activity

Some substances reduce or even stop the catalytic activity of enzymes in biochemical reactions. They block or distort the active site. These chemicals are called inhibitors, because they inhibit reaction. Inhibitors that occupy the active site and prevent a substrate molecule from binding to the enzyme are said to be active site-directed (or competitive, as they 'compete' with the substrate for the active site). Inhibitors that attach to other parts of the enzyme molecule, perhaps distorting its shape, are said to be non-active site-directed (or non competitive).

Immobilized Enzymes

Enzymes are widely used commercially, for example in the detergent, food and brewing industries. Protease enzymes are used in 'biological' washing powders to speed up the breakdown of proteins in stains like blood and egg. Pectinase is used to produce and clarify fruit juices.

Problems using enzymes commercially include:

- They are water soluble which makes them hard to recover
- Some products can inhibit the enzyme activity (feedback inhibition) Enzymes can be immobilized by fixing them to a solid surface. This has a number of commercial advantages:
- The enzyme is easily removed
- The enzyme can be packed into columns and used over a long period
- Speedy separation of products reduces feedback inhibition
- Thermal stability is increased allowing higher temperatures to be used
- Higher operating temperatures increase rate of reaction

There are four principal methods of immobilization currently in use:

- Covalent bonding to a solid support
- Adsorption onto an insoluble substance
- Entrapment within a gel
- Encapsulation behind a selectively permeable membrane

SPECIFICITY OF ENZYMES

One of the properties of enzymes that makes them so important as diagnostic and research tools is the specificity they exhibit relative to the reactions they catalyze. A few enzymes exhibit absolute specificity; that is, they will catalyze only one particular reaction. Other enzymes will be specific for a particular type of chemical bond or functional group.

In general, there are four distinct types of specificity:

1. *Absolute specificity*: The enzyme will catalyze only one reaction.
2. *Group specificity*: The enzyme will act only on molecules that have specific functional groups, such as amino, phosphate and methyl groups.
3. *Linkage specificity*: The enzyme will act on a particular type of chemical bond regardless of the rest of the molecular structure.

4. *Stereochemical specificity*: The enzyme will act on a particular steric or optical isomer.

Though enzymes exhibit great degrees of specificity, cofactors may serve many apoenzymes. For example, nicotinamide adenine dinucleotide (NAD) is a coenzyme for a great number of dehydrogenase reactions in which it acts as a hydrogen acceptor. Among them are the alcohol dehydrogenase, malate dehydrogenase and lactate dehydrogenase reactions.

Naming and Classification

Except for some of the original enzymes such as pepsin, rennin, and trypsin, most enzyme names end in "ase". The International Union of Biochemistry (I.U.B.) initiated standards of enzyme nomenclature which recommend that enzyme names indicate both the substrate acted upon and the type of reaction catalyzed. Under this system, the enzyme uricase is called urate: O_2 oxidoreductase, while the enzyme glutamic oxaloacetic transaminase (GOT) is called L-aspartate: 2-oxoglutarate aminotransferase.

Enzymes can be classified by the kind of chemical reaction catalyzed:

- Addition or removal of water
  - *Hydrolases*: These include esterases, carbohydrases, nucleases, deaminases, amidases, and proteases
  - Hydrases such as fumarase, enolase, aconitase and carbonic anhydrase
- Transfer of electrons
  - Oxidases
  - Dehydrogenases
- Transfer of a radical
  - *Transglycosidases*: Of monosaccharides
  - *Transphosphorylases and phosphomutases*: Of a phosphate group
  - *Transaminases*: Of amino group
  - *Transmethylases*: Of a methyl group
  - *Transacetylases*: Of an acetyl group
- Splitting or forming a C-C bond
  - Desmolases
- Changing geometry or structure of a molecule
  - Isomerases

- Joining two molecules through hydrolysis of pyrophosphate bond in ATP or other triphosphate
  - Ligases

TYPES OF ENZYMES

- Simple enzymes are those composed completely of proteins
- Complex enzymes are those composed of protein and a small organic molecule (s) (also known as holoenzymes)

Enzymes are Categorized by their Function (s):

- Metabolic enzymes catalyze and regulate every biochemical reaction that occurs in your body and are essential to cellular function and health.
- Digestive enzymes are secreted along the digestive tract, break down food into nutrients and waste, and allow the nutrients found in foods to be absorbed into the bloodstream. The pancreas produces most digestive enzymes, but the liver, gallbladder, small intestines, stomach and colon also play critical roles in the production of these enzymes.
  Examples of digestive enzymes: lipase, protease, amylase, ptyalin, pepsin and trypsin.
- Food enzymes enter the body through the consumption of raw foods or enzyme supplements. Cooking and processing destroys most enzymes in food.
- Plant Based enzymes are synthetically grown, help break down fat, protein and carbohydrates and function within a broad pH range. Plant enzymes, commonly bromelin (pineapple) and papain (papaya) play an important role in more complete digestion of all foods. They are activated at a temperature higher than normal body temperature, also making them a good anti-inflammatory.
- Proteoplytic enzymes are enzymes that digest proteins, including Trypsin, Chymotrypsin, Pancreatin, Bromelin and Papain. Some are produced by the pancreas and others are supplemental from animals or plants. The primary use of proteolytic enzymes in supplements is, are as digestive aids. They are also used in drugs as anti-inflammatory agents and pain relievers.

ACTIVE SITES AND CATALYTIC MECHANISMS

Open any textbook of biochemistry and you will be presented with an overwhelming number of figures depicting the crystal structures of a multitude of enzymes. Of course, the three-dimensional structures of enzymes are crucial to their functions.

Many enzymes are just regular protein molecules, composed of nothing else than the 20 standard amino acids. If you mix of all these amino acids in free form at, say, 10 mM each, this mixture will not have any significant catalytic activity. It therefore is the precise arrangement of the amino acids in the enzyme molecule that brings about the function.

As the á-amino and á-carboxyl groups of the amino acids are hooked up to each other in the polypeptide chain, they usually do not directly contribute to the catalytic effect of the enzyme. Instead, it is the side chains that are directly engaged win the reaction. A very good example of this is chymotrypsin. Chymotrypsin is one of the major proteases in the human digestive tract, where its job is to knock down large protein molecules into small peptides that are then further processed by peptidases.

How does chymotrypsin do that? The enzyme-catalyzed reaction is similar to alkaline hydrolysis. In alkaline hydrolysis, a hydroxide ion, which is a strong nucleophile, attacks the carbon in the peptide bond that carries a partial positive charge. In the enzyme-catalyzed reaction, a deprotonated serine residue of the enzyme plays a role similar to that of the hydroxide ion.

Now, you know that serine normally is a neutral amino acid – its OH-group does not spontaneously dissociate, no more than the-OH group of alcohol does. The question therefore is, how does the enzyme deprotonate its own serine side chain? This is brought about by placing the serine next to a histidine and an aspartate inside the active centre.

Aspartate deprotonates histidine, which in turn deprotonates the serine residue. This motive – asp, his, ser – is very widespread among proteases and esterases, so much so that it is commonly called 'the catalytic triad'. E.g., the protease trypsin and several lipases that occur in human metabolism have this motif and share the same mechanism of catalysis.

Structure and Mechanism of Chymotrypsin:

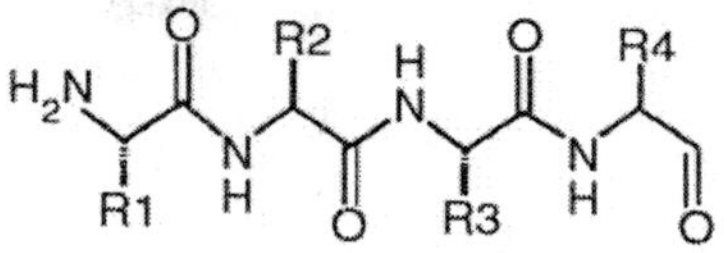

Fig. The Peptide Bond cleaved by chymotrypsin.

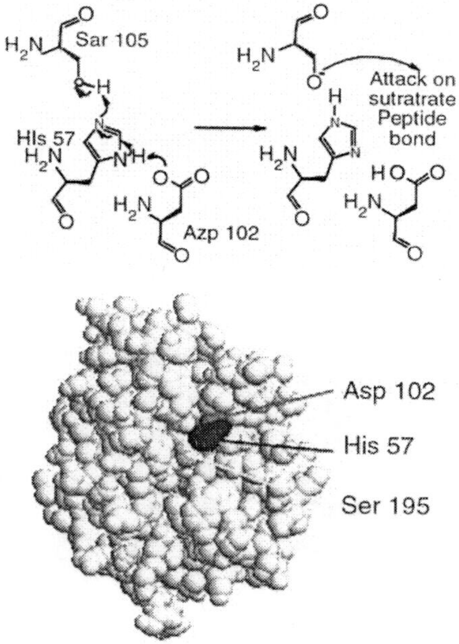

Fig. The Interplay of the Side Chains of Histidine 57, Aspartate 102, and Serine 195 that Results in the Deprotonation of Serine.

All three side chains are close to each other in the active site of the enzyme. While with many enzymes the protein molecule and its amino acid side chains are sufficient for catalysis, many others require co-enzymes for their catalytic activity. Very often, both active-site amino acid side chains and coenzymes are required. For example, amino acid transaminases have a molecule of the coenzyme pyridoxalphosphate bound to the active site, which co-operates with an active site lysine in the enzyme's catalytic activity. Most enzymes have just one active site, or if they are multimeric one active site per subunit. However, there are exceptions: Fatty acid synthase has as many as eight different active sites on each subunit. Multi-enzyme complexes such as pyruvate dehydrogenase have one active site per subunit but combine different types of subunits and enzyme activities in one functional assembly.

Fig. The Enzyme-Catalized reaction.

Fig. The Base-Catalyzed Reaction for Comparison.

EFFECTS OF INHIBITORS ON ENZYME ACTIVITY

Enzyme inhibitors are substances which alter the catalytic action of the enzyme and consequently slow down, or in some cases, stop catalysis. There are three common types of enzyme inhibition-competitive, non-competitive and substrate inhibition.

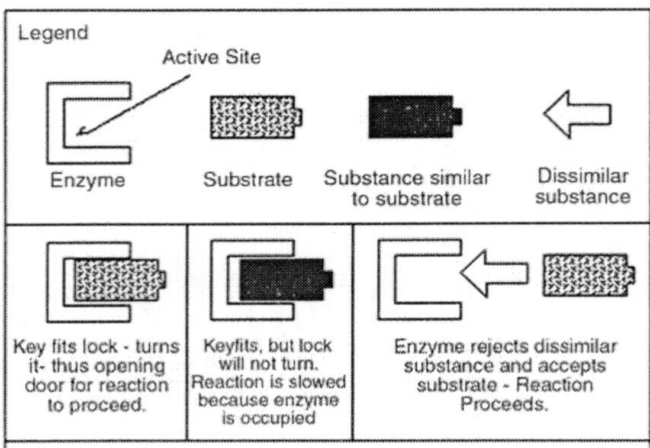

Fig. Lock-key Theory-Competitive Inhibition

Most theories concerning inhibition mechanisms are based on the existence of the enzyme-substrate complex ES. As mentioned earlier, the existence of temporary ES structures has been verified in the laboratory.

Competitive inhibition occurs when the substrate and a substance resembling the substrate are both added to the enzyme. A theory called the "lock-key theory" of enzyme catalysts can be used to explain why inhibition occurs.

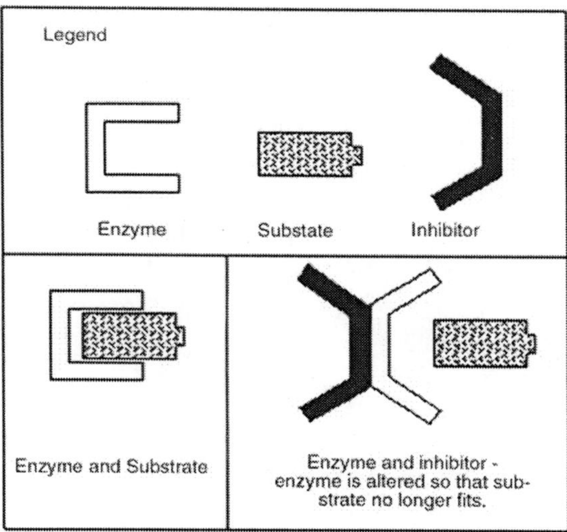

Fig. Non-Competitive Inhibition

The lock and key theory utilizes the concept of an "active site." The concept holds that one particular portion of the enzyme surface has a strong affinity for the substrate. The substrate is held in such a way that its conversion to the reaction products is more favorable. If we consider the enzyme as the lock and the substrate the key-the key is inserted in the lock, is turned, and the door is opened and the reaction proceeds.

However, when an inhibitor which resembles the substrate is present, it will compete with the substrate for the position in the enzyme lock. When the inhibitor wins, it gains the lock position but is unable to open the lock. Hence, the observed reaction is slowed down because some of the available enzyme sites are occupied by the inhibitor.

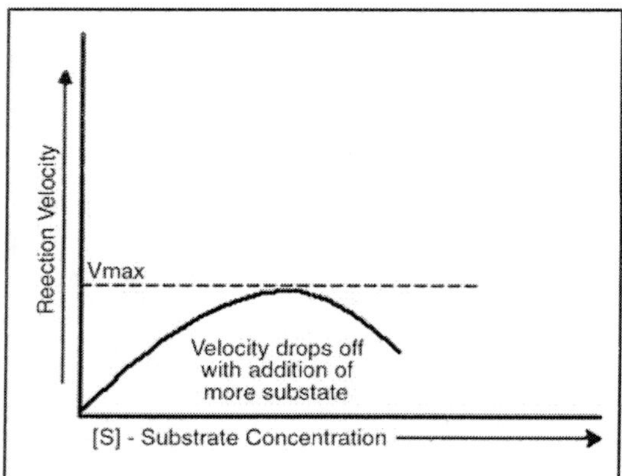

Fig. Substrate Become Rate-inhibiting

If a dissimilar substance which does not fit the site is present, the enzyme rejects it, accepts the substrate, and the reaction proceeds normally. Non-competitive inhibitors are considered to be substances which when added to the enzyme alter the enzyme in a way that it cannot accept the substrate. Substrate inhibition will sometimes occur when excessive amounts of substrate are present. Figure shows the reaction velocity decreasing after the maximum velocity has been reached. Additional amounts of substrate

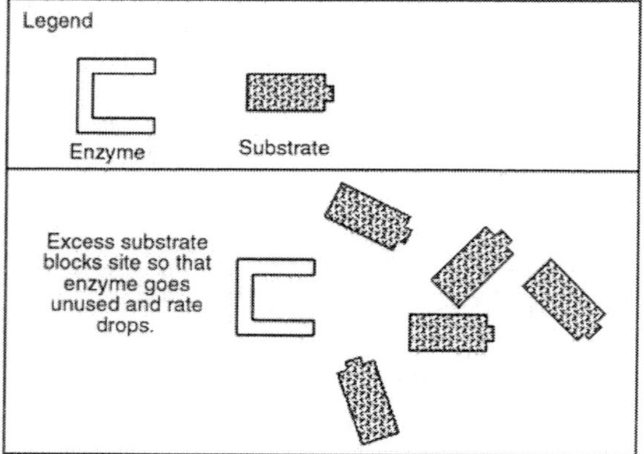

Fig. Substrate Inhibition

added to the reaction mixture after this point actually decrease the reaction rate. This is thought to be due to the fact that there are so many substrate molecules competing for the active sites on the enzyme surfaces that they block the sites and prevent any other substrate molecules from occupying them. This causes the reaction rate to drop since all of the enzyme present is not being used.

Temperature Effects

Like most chemical reactions, the rate of an enzyme-catalyzed reaction increases as the temperature is raised. A ten degree Centigrade rise in temperature will increase the activity of most enzymes by 50 to 100%. Variations in reaction temperature as small as 1 or 2 degrees may introduce changes of 10 to 20% in the results. In the case of enzymatic reactions, this is complicated by the fact that many enzymes are adversely affected by high temperatures. The reaction rate increases with temperature to a maximum level, then abruptly declines with further increase of temperature. Because most animal enzymes rapidly become denatured at temperatures above 40°C, most enzyme determinations are carried out somewhat below that temperature. Over a period of time, enzymes will be deactivated at even moderate temperatures. Storage of enzymes at 5°C or below is generally the most suitable. Some enzymes lose their activity when frozen.

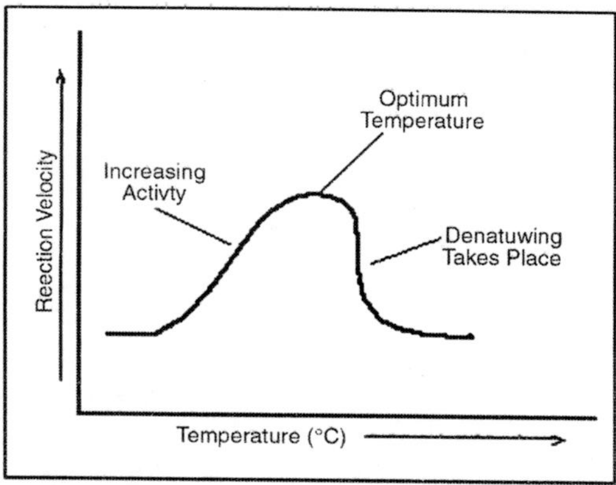

Fig. Effect of Temperature on Reaction Rate

Effects of pH

Enzymes are affected by changes in pH. The most favorable pH value-the point where the enzyme is most active-is known as the optimum pH.

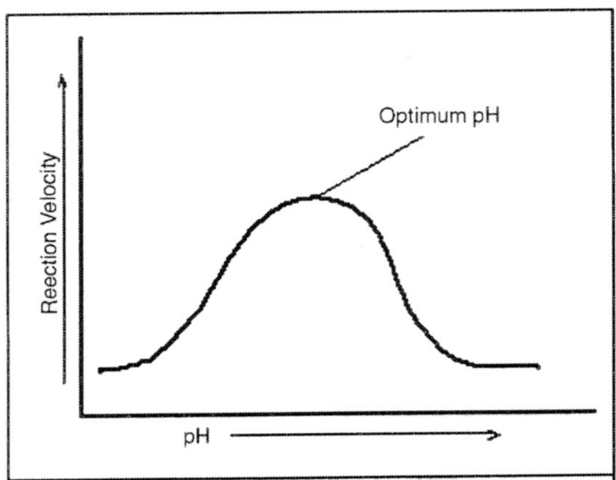

Extremely high or low pH values generally result in complete loss of activity for most enzymes. pH is also a factor in the stability of enzymes. As with activity, for each enzyme there is also a region of pH optimal stability. The optimum pH value will vary greatly from one enzyme to another, table shows: In addition to temperature and pH there are other factors, such as ionic strength, which can affect the enzymatic reaction. Each of these physical and chemical parameters must be considered and optimized in order for an enzymatic reaction to be accurate and reproducible.

Table. pH for Optimum Activity.

Enzyme	pH Optimum
Lipase (pancreas)	8.0
Lipase (stomach)	4.0-5.0
Lipase (castor oil)	4.7
Pepsin	1.5-1.6
Trypsin	7.8-8.7
Urease	7.0
Invertase	4.5
Maltase	6.1-6.8
Amylase (pancreas)	6.7-7.0
Amylase (malt)	4.6-5.2
Catalase	7.0

The term enzyme kinetics implies a study of the speed, rate or velocity of an enzyme catalysed reaction, and of the various factors which may affect this. At the heart of any study of enzyme kinetics is a knowledge of the way in which reaction velocity is altered by changes in the concentration of the enzyme's substrate and of the simple mathematics underlying this. To ease ourselves gently into this we will assume that the enzyme that we are discussing has no special features, such as allosteric properties, and catalyses the conversion of just one substrate to one product. This may seem unrelated to real enzymes, very few have just one substrate after all, but it will provide a basis which we can expand on later when we study more complex systems.

Enzyme kinetics is the study of the chemical reactions that are catalysed by enzymes, with a focus on their reaction rates. The study of an enzyme's kinetics reveals the catalytic mechanism of this enzyme, its role in metabolism, how its activity is controlled, and how a drug or a poison might inhibit the enzyme. Enzymes are usually protein molecules that manipulate other molecules — the enzymes' substrates. These target molecules bind to an enzyme's active site and are transformed into products through a series of steps known as the enzymatic mechanism. These mechanisms can be divided into single-substrate and multiple-substrate mechanisms. Kinetic studies on enzymes that only bind one substrate, such as triosephosphate isomerase, aim to measure the affinity with which the enzyme binds this substrate and the turnover rate.

When enzymes bind multiple substrates, such as dihydrofolate reductase (shown right), enzyme kinetics can also show the sequence in which these substrates bind and the sequence in which products are released. An example of enzymes that bind a single substrate and release multiple products are proteases, which cleave one protein substrate into two polypeptide products. Others join two substrates together, such as DNA polymerase linking a nucleotide to DNA. Although these mechanisms are often a complex series of steps, there is typically one rate-determining step that determines the overall kinetics. This rate-determining step may be a chemical reaction or a conformational change of the enzyme or substrates, such as those involved in the release of product(s) from the enzyme.

Knowledge of the enzyme's structure is helpful in interpreting the kinetic data. For example, the structure can suggest how substrates and products bind during catalysis; what changes occur during the reaction; and even the role of particular amino acid residues in the mechanism. Some enzymes change shape significantly during the mechanism; in such cases, it is helpful to determine the enzyme structure with and without bound substrate analogs that do not undergo the enzymatic reaction. Not all biological catalysts are protein enzymes; RNA-based catalysts such as ribozymes and ribosomes are essential to many cellular functions, such as RNA splicing and translation. The main difference between ribozymes and enzymes is that the RNA catalysts perform a more limited set of reactions, although their reaction mechanisms and kinetics can be analysed and classified by the same methods.

The reaction catalysed by an enzyme uses exactly the same reactants and produces exactly the same products as the uncatalysed reaction. Like other catalysts, enzymes do not alter the position of equilibrium between substrates and products. However, unlike normal chemical reactions, enzymes are saturable. This means as more substrate is added, the reaction rate will increase, because more active sites become occupied. This can continue until all the enzyme becomes saturated with substrate and the rate reaches a maximum. The two most important kinetic properties of an enzyme are how quickly the enzyme becomes saturated with a particular substrate, and the maximum rate it can achieve. Knowing these properties suggests what an enzyme might do in the cell and can show how the enzyme will respond to changes in these conditions.

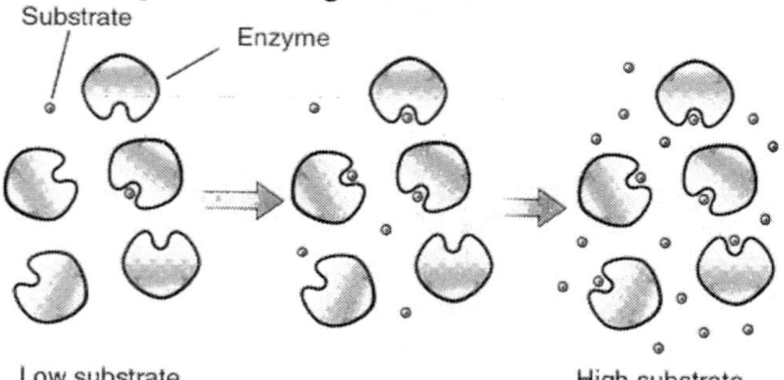

Fig. Reaction rates Increase as Substrate Concentration Increase, but Become Saturated at very High Concentrations of Substrate.

Enzyme assays are laboratory procedures that measure the rate of enzyme reactions. Because enzymes are not consumed by the reactions they catalyse, enzyme assays usually follow changes in the concentration of either substrates or products to measure the rate of reaction. There are many methods of measurement. Spectrophotometric assays observe change in the absorbance of light between products and reactants; radiometric assays involve the incorporation or release of radioactivity to measure the amount of product made over time. Spectrophotometric assays are most convenient since they allow the rate of the reaction to be measured continuously. Although radiometric assays require the removal and counting of samples (i. e., they are discontinuous assays) they are usually extremely sensitive and can measure very low levels of enzyme activity. An analogous approach is to use mass spectrometry to monitor the incorporation or release of stable isotopes as substrate is converted into product. The most sensitive enzyme assays use lasers focused through a microscope to observe changes in single enzyme molecules as they catalyse their reactions.

These measurements either use changes in the fluorescence of cofactors during an enzyme's reaction mechanism, or of fluorescent dyes added onto specific sites of the protein to report movements that occur during catalysis. These studies are providing a new view of the kinetics and dynamics of single enzymes, as opposed to traditional enzyme kinetics, which observes the average behaviour of populations of millions of enzyme molecules.

On the left is shown a typical progress curve for an enzyme assay. The enzyme produces product at a linear initial rate at the start of the reaction. Later in this progress curve, the rate slows down as substrate is used up or products accumulate. The length of the initial rate period depends on the assay conditions and can range from milliseconds to hours. Enzyme assays are usually set up to produce an initial rate lasting over a minute, to make measurements easier. However, equipment for rapidly mixing liquids allows fast kinetic measurements on initial rates of less than one second. These very rapid assays are essential for measuring pre-steady-state kinetics.

Most enzyme kinetics studies concentrate on this initial, linear part of enzyme reactions. However, it is also possible to measure the complete reaction curve and fit this data to a non-linear rate equation. This way of measuring enzyme reactions is called progress-curve analysis. This approach is useful as an alternative to rapid kinetics when the initial rate is too fast to measure accurately.

WORKING PROCESS OF ENZYMES

TYPES OF ENZYMATIC REACTIONS

When looking at enzyme names such as 'transketolase' or 'phosphorylase', you will note that these names don't tell you exactly what reactions the enzymes may catalyze. A complete description should mention the coenzymes required, the substrates and the particular bonds in the substrates that are being severed or created.

A nomenclature that meets these criteria has been developed by the Enzyme Commission of the IUBMB (International Union of Biochemistry and Molecular Biology). In the IUBMB nomenclature, the enzyme transketolase bears the formidable name:

Sedoheptulose-7-phosphate: D-glyceraldehyde-3-phosphate glycolaldehydetransferase. Such names, of course, are rather lengthy, and their use is not very widespread. To make the tasks of tracking and bookkeeping more manageable, these names are supplemented with numeric codes. In the IUBMB scheme, enzymes are put into one out of six classes according to the reactions they catalyze.

These classes are:

- *Oxidoreductases*. These catalyze redox reactions, frequently involving one of the coenzymes NAD^+, $NADP^+$, or FAD.
- *Transferases*. These bring about the transfer of functional groups – e.g., phosphate groups from ATP to another metabolite, which activates the latter and sets it up for subsequent reaction steps.
- *Hydrolases*. These catalyze hydrolysis reactions – e.g., such as those involved in digestion of foodstuffs.
- *Lyases* – these effect elimination reactions that result in the formation of double bonds.
- *Isomerases*. These facilitate the interconversion of isomers. We will meet two examples as soon as we get into glycolysis.
- Ligases, which form new covalent bonds at the expense of ATP hydrolysis.

Of course, within each of these main classes, there are subclasses and sub-sub classes that correspond to details of substrates and mechanisms of the enzyme reactions. Each individual enzyme activity is assigned an individual number within a sub-sub class, so that we wind up with a four-figure designation, which is preceded by the letters 'EC' (Enzyme Commission). One good thing about this classification is that it is rarely used – the 'recommended names', which most of the time happen to be the traditional ones, are used instead. The other good thing is that it has a sound appreciation of priorities. The single most important enzyme in student lifestyle – namely, alcohol dehydrogenase (or, as IUBMB puts it, alcohol:NAD oxidoreductase). This beneficial enzyme, residing in the liver, degrades ethanol, and without it you would be drunk all the time!

Enzyme Kinetics: Basic Enzyme Reactions

Enzymes are catalysts and increase the speed of a chemical reaction without themselves undergoing any permanent chemical change. They are neither used up in the reaction nor do they appear as reaction products. The basic enzymatic reaction can be represented as follows where E represents the enzyme catalyzing the reaction, S the substrate, the substance being changed, and P the product of the reaction.

Enzyme Kinetics: Energy Levels

Chemists have known for almost a century that for most chemical reactions to proceed, some form of energy is needed. They have termed this quantity of energy, "the energy of activation." It is the magnitude of the activation energy which determines just how fast the reaction will proceed. It is believed that enzymes lower the activation energy for the reaction they are catalyzing. Figure illustrates this concept.

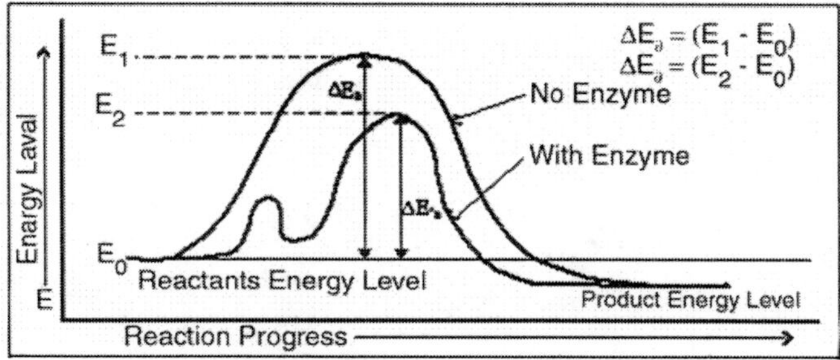

Fig. Energy of Activation Δ_∂ E is less than Δ_∂ E due to the Effect of the Enzyme on the Substrate.

The enzyme is thought to reduce the "path" of the reaction. This shortened path would require less energy for each molecule of substrate converted to product. Given a total amount of available energy, more molecules of substrate would be converted when the enzyme is present (the shortened "path") than when it is absent. Hence, the reaction is said to go faster in a given period of time.

ENZYME KINETICS

A theory to explain the catalytic action of enzymes was proposed by the Swedish chemist Savante Arrhenius in 1888. He proposed that the substrate and enzyme formed some intermediate substance which is known as the enzyme substrate complex.

The reaction can be represented as:

$$\underset{\text{substrate}}{S} + \underset{\text{enzyme}}{E} \longrightarrow \underset{\text{enzyme substrate complex}}{ES}$$

If this reaction is combined with the original reaction equation, the following results:

$$\underset{\text{substrate}}{S} + \underset{\text{enzyme}}{E} \longrightarrow \underset{\text{enzyme substrate complex}}{ES} \longrightarrow \underset{\text{Product}}{P} + \underset{\text{enzyme}}{E}$$

The existence of an intermediate enzyme-substrate complex has been demonstrated in the laboratory, for example, using catalase and a hydrogen peroxide derivative. At Yale University, Kurt G. Stern observed spectral shifts in catalase as the reaction it catalyzed proceeded. This experimental evidence indicates that the enzyme first unites in some way with the substrate and then returns to its original form after the reaction is concluded.

Chemical Equilibrium

The study of a large number of chemical reactions reveals that most do not go to true completion. This is likewise true of enzymatically-catalyzed reactions.

This is due to the reversibility of most reactions. In general:

$$A + B \xrightarrow{K+1} C + D \quad \text{forward reaction}$$

$$C + D \xrightarrow{K-1} A + B \quad \text{reverse reaction}$$

where K^{+1} is the forward reaction rate constant and K^{-1} is the rate constant for the reverse reaction.

Combining the two reactions gives:

$$A + B \underset{K-1}{\overset{K+1}{\rightleftarrows}} C + D$$

Applying this general relationship to enzymatic reactions allows the equation:

$$E + S \underset{K-1}{\overset{K+1}{\rightleftarrows}} ES \underset{K-2}{\overset{K+2}{\rightleftarrows}} p + E$$

Equilbrium, a steady state condition, is reached when the forward reaction rates equal the backward rates.

This is the basic equation upon which most enzyme activity studies are based.

FACTORS AFFECTING ENZYME ACTIVITY

Knowledge of basic enzyme kinetic theory is important in enzyme analysis in order both to understand the basic enzymatic mechanism and to select a method for enzyme analysis. The conditions selected to measure the activity of an enzyme would not be the same as those selected to measure the concentration of its substrate.

Several factors affect the rate at which enzymatic reactions proceed-temperature, pH, enzyme concentration, substrate concentration, and the presence of any inhibitors or activators.

Enzyme Concentration

In order to study the effect of increasing the enzyme concentration upon the reaction rate,the substrate must be present in an excess amount; i.e., the reaction must be independent of the substrate concentration.

Any change in the amount of product formed over a specified period of time will be dependent upon the level of enzyme present. These reactions are said to be "zero order" because the rates are independent of substrate concentration, and are equal to some constant k.

The formation of product proceeds at a rate which is linear with time. The addition of more substrate does

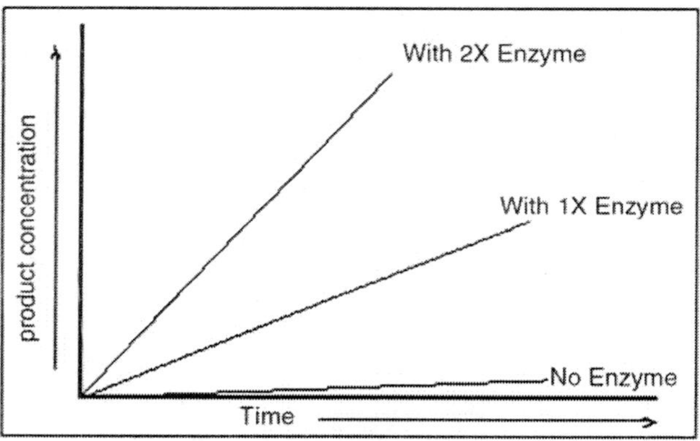

Fig. "Zer order" Reaction Rate is Independent of Substrate Concentration.

not serve to increase the rate. In zero order kinetics, allowing the assay to run for double time results in double the amount of product.

Table. Reaction Orders with Respect to Substrate Concentration

Order	Rate Equation	Comments
Zero	rate = k	rate is independent of substrate concentration
First	rate = k[S]	rate is proportional to the first power of substrate concentration
Second	rate = $k[S][S]=k[S]^2$	rate is proportional to the square of the substrate concentration
Second	rate = $k[S_1][S_2]$	rate is proportional to the first power of each of two reactants

The amount of enzyme present in a reaction is measured by the activity it catalyzes. The relationship between activity and concentration is affected by many factors such as temperature, pH, etc. An enzyme assay must be designed so that the observed activity is proportional to the amount of enzyme present in order that the enzyme concentration is the only limiting factor. It is satisfied only when the reaction is zero order. In Figure, activity is directly proportional to concentration in the area AB, but not in BC. Enzyme activity is generally greatest when substrate concentration is unlimiting. When the concentration of the product of an enzymatic Reaction is plotted against time, a similar curve results. Between A and B, the curve represents a zero order reaction; that is, one in which the rate is constant with time.

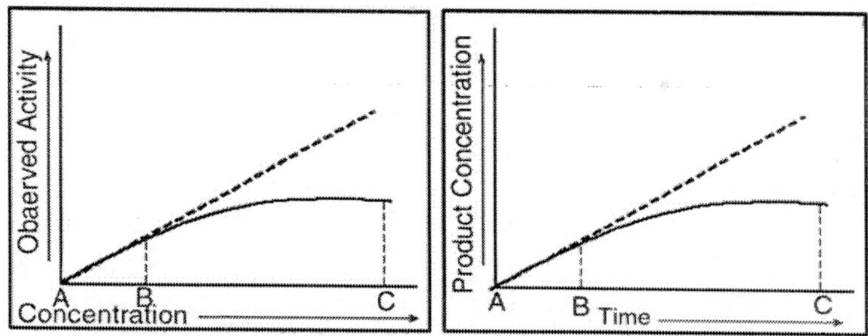

Fig. (a) Activity Concentration. (b) Reaction rate limited by substrate concentration

As substrate is used up, the enzyme's active sites are no longer saturated, substrate concentration becomes rate limiting, and the reaction becomes first order between B and C. To measure enzyme activity ideally, the measurements must be made in that portion of the curve where the reaction is zero order.

A reaction is most likely to be zero order initially since substrate concentration is then highest. To be certain that a reaction is zero order, multiple measurements of product (or substrate) concentration must be made.

Three types of reactions which might be encountered in enzyme assays and shows the problems which might be enountered if only single measurements are made.

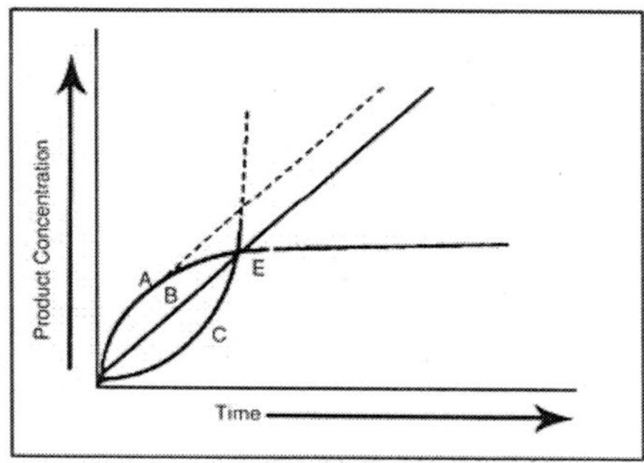

Fig. Leading, Lagging and Liner Reactions

B is a straight line representing a zero order reaction which permits accurate determination of enzyme activity for part or all of the reaction time. This reaction is zero order initially and then slows, presumably due to substrate exhaustion or product inhibition. This type of reaction is sometimes referred to as a "leading" reaction. True "potential" activity is represented by the dotted line. Curve C represents a reaction with an initial "lag" phase. Again the dotted line represents the potentially measurable activity. Multiple determinations of product concentration enable each curve to be plotted and true activity determined. A single end point determination at E would lead to the false conclusion that all three samples had identical enzyme concentration.

Substrate Concentration

It has been shown experimentally that if the amount of the enzyme is kept constant and the substrate concentration is then gradually increased, the reaction velocity will increase until it reaches a maximum. After this point, increases in substrate concentration will not increase the velocity (Δ A/ Δ).

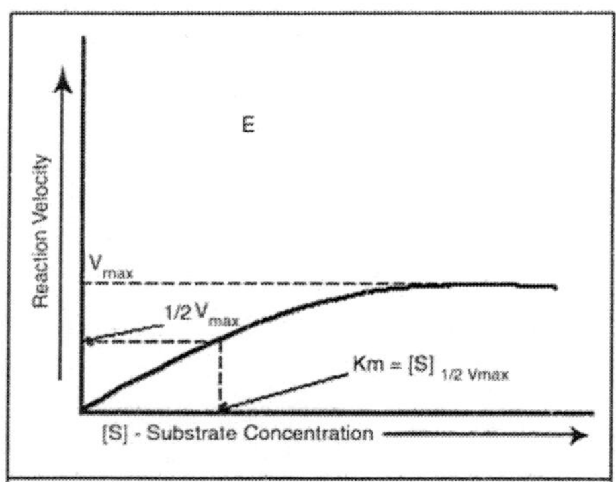

Fig. Effect of SConcentration

It is theorized that when this maximum velocity had been reached, all of the available enzyme has been converted to ES, the enzyme substrate complex. This point on the graph is designated V_{max}. Using this maximum velocity and equation, Michaelis developed a set of mathematical expressions to calculate enzyme activity in terms of reaction speed from measurable laboratory data.

$$E + S \underset{K-1}{\overset{K+1}{\rightleftharpoons}} ES \underset{K-2}{\overset{K+2}{\rightleftharpoons}} p + E$$

The Michaelis constant Km is defined as the substrate concentration at 1/2 the maximum velocity. Using this constant and the fact that Km can also be defined as:

$$K_m = K_{-1} + K_{2/}K_{+1}$$

K^{+1}, K^{-1} and K^{+2} being the rate constants from equation. Michaelis developed the following:

$$V_1 \frac{V_{mx}[S]}{K_m + [S]}$$

V_1 = The velocity at any time

[S] = The substrate concentration at this time

V_{mx} = The highest under this set of experimental condition (pH, temperature, etc.)

K_M = The Michaselis constant for the particular enzyme being investigated Michaelis constants have been determined for many of the commonly used enzymes. The size of Km tells us several things about a particular enzyme.

- A small Km indicates that the enzyme requires only a small amount of substrate to become saturated. Hence, the maximum velocity is reached at relatively low substrate concentrations.
- A large Km indicates the need for high substrate concentrations to achieve maximum reaction velocity.
- The substrate with the lowest Km upon which the enzyme acts as a catalyst is frequently assumed to be enzyme's natural substrate, though this is not true for all enzymes.

Chapter 6

Separation in Food Processing

Biological raw materials are usually mixtures, and to prepare foodstuffs it may be necessary to separate some of the components of the mixtures. One method, by which this separation can be carried out, is by the introduction of a new phase to the system and allowing the components of the original raw material to distribute themselves between the phases.

For example, freshly dug vegetables have another phase, water, added to remove unwanted earth; a mixture of alcohol and water is heated to produce another phase, vapour, which is richer in alcohol than the mixture. By choosing the conditions, one phase is enriched whilst the other is depleted in some component or components.

The maximum separation is reached at the equilibrium distribution of the components, but in practice separation may fall short of this as equilibrium is not attained. The components are distributed between the phases in accordance with equilibrium distribution coefficients which give the relative concentrations in each phase when equilibrium has been reached. The two phases can then be separated by simple physical methods such as gravity settling.

This process of contact, redistribution, and separation gives the name contact equilibrium separations. Successive stages can be used to enhance the separation. An example is in the extraction of edible oil from soya beans. Beans containing oil are crushed, and then mixed with a solvent in which the oil, but not the other components of the beans, is soluble. Initially, the oil will be distributed between the beans and the solvent, but after efficient crushing and mixing the oil will be dissolved in the solvent. In the separation, some solvent and oil will be retained by the mass of beans; these will constitute one stream and the bulk of the solvent and oil the other. This process of contacting the two streams, of crushed beans and solvent, makes up one contact stage. To extract more oil from the beans, further contact stages can be provided by mixing the extracted beans with a fresh stream of solvent.

For economy and convenience, the solvent and oil stream from another extraction is often used instead of fresh solvent. So two streams, one containing beans and the other starting off as pure solvent, can move counter current to each other through a series of contact stages with progressive contacting followed by draining. In each stage of the process in which the streams come into contact, the material being transferred is distributed in equilibrium between the two streams. By removing the streams from the contact stage and contacting each with material of different composition, new equilibrium conditions are established and so separation can proceed.

In order to effect the desired separation of oil from beans, the process itself has introduced a further separation problem—the separation of the oil from the solvent. However, the solvent is chosen so that this subsequent separation is simple, for example, by distillation. In some cases, such as washing, further separation of dissolved material from wash water may not be necessary and one stream may be rejected as waste. In other cases, such as distillation, the two streams are generated from the mixture of original components by vapourization of part of the mixture.

The two features that are common to all equilibrium contact processes are the attainment of, or approach to, equilibrium and the provision of contact stages. Equilibrium is reached when a component is so distributed between the two streams that there is no tendency for its concentration in either stream to change. Attainment of equilibrium may take appreciable time, and only if this time is available will effective equilibrium be reached. The opportunity to reach equilibrium is provided in each stage, and so with one or more stages the concentration of the transferred component changes progressively from one stream to the other, providing the desired separation.

MEMBRANE SEPARATIONS

Membranes can be used for separating constituents of foods on a molecular basis, where the foods are in solution and where a solution is separated from one less concentrated by a semi-permeable membrane. These membranes act somewhat as membranes do in natural biological systems.

Water flows through the membrane from the dilute solution to the more concentrated one. The force producing this flow is called the osmotic pressure and to stop the flow a pressure, equal to the osmotic pressure, has to be exerted externally on the more concentrated solution. Osmotic pressures in liquids arise in the same way as partial pressures in gases: using the number of moles of the solute present and the volume of the whole solution, the osmotic pressure can be estimated using the gas laws. If pressures greater than the osmotic pressure are applied to the more concentrated solution, the flow will not only stop but will reverse so that water passes out through the membrane making the concentrated solution more concentrated.

The flow will continue until the concentration rises to the point where its osmotic pressure equals the applied pressure. Such a process is called reverse osmosis and special artificial membranes have been made with the required "tight" structure to retain all but the smallest molecules such as those of water.

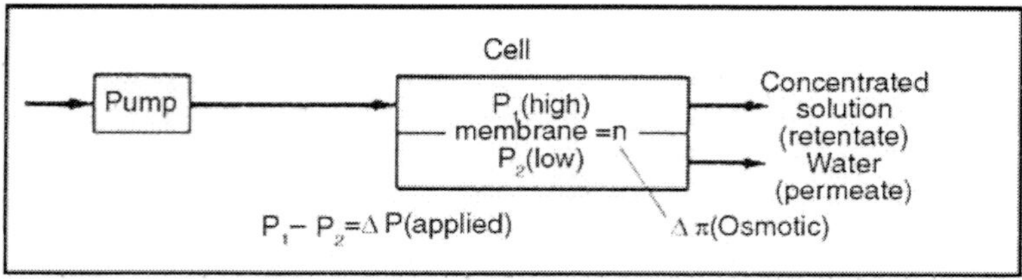

Fig. Reverse Osmosis Systems

Also, "looser" membranes have been developed through which not only water, but also larger solute molecules can selectively pass if driven by imposed pressure. Membranes are available which can retain large molecules, such as proteins, while allowing through smaller molecules.

Because the larger molecules are normally at low molar concentrations, they exert very small osmotic pressures, which therefore enter hardly at all into the situation. Following the resemblance to conventional filtration, this process is called *ultrafiltration.*

In general, ultrafiltration needs relatively low differential pressures, up to a few atmospheres. If higher pressures are used, a protein or solute gel appears to form on the membrane, which resists flow, and so the increased pressure may not increase the transfer rate. Important applications for ultrafiltration are for concentrating solutions of large polymeric molecules, such as milk and blood proteins.

Another significant application is to the concentration of whey proteins. Reverse osmosis, on the other hand, is concerned mainly with solutions containing smaller molecules such as simple sugars and salts at higher molar concentrations, which exert higher osmotic pressures.

To overcome these osmotic pressures, high external pressures have to be exerted, up to the order of 100 atmospheres. Limitations to increased flow rates arise in this case from the mechanical weaknesses of the membrane and from concentration of solutes which causes substantial osmotic "back" pressure. Applications in the food industry are in separating water from, and thus concentrating, solutions such as fruit juices.

Membrane Equipment

The equipment for these membrane separation processes consists of the necessary pumps, flow systems and membranes. In the case of ultrafiltration, the membranes are set up in a wide variety of geometrical arrangements, mostly tubular but sometimes in plates, which can be mounted similarly to a filter press or plate heat exchanger. Flow rates are kept high over the surfaces and recirculation of the fluid on the high pressure, or retentate, side is often used; the fluid passing through, called the permeate, is usually collected in suitable troughs or tanks at atmospheric pressure.

In the case of reverse osmosis, the high pressures dictate mechanical strength, and stacks of flat disc membranes can be used one above the other. Another system uses very small diameter (around 0.04 mm) hollow filaments on plastic supports; the diametres are small to provide strength but preclude many food solutions because of this very small size. The main flow in reverse osmosis is the permeate. The systems can be designed either as continuous or as batch operations. One limitation to extended operation arises from the need to control growth of bacteria. After a time bacterial concentrations in the system, for example, in the gel at the surface of the ultrafiltration membranes, can grow so high that cleaning must be provided. This can be difficult as many of the membranes are not very robust either to mechanical disturbance or to the extremes of pH which could give quicker and better cleaning.

Distillation

Distillation is a separation process, separating components in a mixture by making use of the fact that some components vapourize more readily than others. When vapours are produced from a mixture, they contain the components of the original mixture, but in proportions which are determined by the relative volatilities of these components. The vapour is richer in some components, those that are more volatile, and so a separation occurs.

In fractional distillation, the vapour is condensed and then re—evapourated when a further separation occurs. It is difficult and sometimes impossible to prepare pure components in this way, but a degree of separation can easily be attained if the volatilities are reasonably different. Where great purity is required, successive distillations may be used.

Major uses of distillation in the food industry are for concentrating essential oils, flavours and alcoholic beverages, and in the deodorization of fats and oils. The equilibrium relationships in distillation are governed by the relative vapour pressures of the mixture components, that is by their volatility relative to one another.

The equilibrium curves for two-component vapour-liquid mixtures can conveniently be presented in two forms, as boiling temperature/concentration curves, or as vapour/liquid concentration distribution curves. Both forms are related as they contain the same data and the concentration distribution curves, which are much the same as the equilibrium curves used in extraction, can readily be obtained from the boiling temperature/concentration curves.

Steam Distillation

In some circumstances in the food industry, distillation would appear to be a good separation method but it cannot be employed directly as the distilling temperatures would lead to breakdown of the materials. In cases in which volatile materials have to be removed from relatively non-volatile materials, steam distillation may sometimes be used to effect the separation at safe temperatures.

A liquid boils when the total vapour pressure of the liquid is equal to the external pressure on the system. Therefore, boiling temperatures can be reduced by reducing the pressure on the system; for example, by boiling under a vacuum, or by adding an inert vapour which by contributing to the vapour pressure, allows the liquid to boil at a lower temperature. Such an addition must be easily removed from the distillate, if it is unwanted in the product, and it must not react with any of the components that are required as products. The vapour that is added is generally steam and the distillation is then spoken of as steam distillation.

Vacuum Distillation

Reduction of the total pressure in the distillation column provides another means of distilling at lower temperatures. When the vapour pressure of the volatile substance reaches the system pressure, distillation occurs. With modern efficient vacuum-producing equipment, vacuum distillation is tending to supplant steam distillation. In some instances, the two methods are combined in vacuum steam distillation.

Batch Distillation

Batch distillation is the term applied to equipment into which the raw liquid mixture is admitted and then boiled for a time. The vapours are condensed. At the end of the distillation time, the liquid remaining in the still is withdrawn as the residue. In some cases the distillation is continued until the boiling point reaches some predetermined level, thus separating a volatile component from a less volatile residue. In other cases, two or more fractions can be withdrawn at different times and these will be of decreasing volatility. During batch distillation, the concentrations change both in the liquid and in the vapour.

Distillation Equipment

The conventional distillation equipment for the continuous fractionation of liquids consists of three main items: a boiler in which the necessary heat to vapourize the liquid is supplied, a column in which the actual contact stages for the distillation separation are provided, and a condenser for condensation of the final top product.

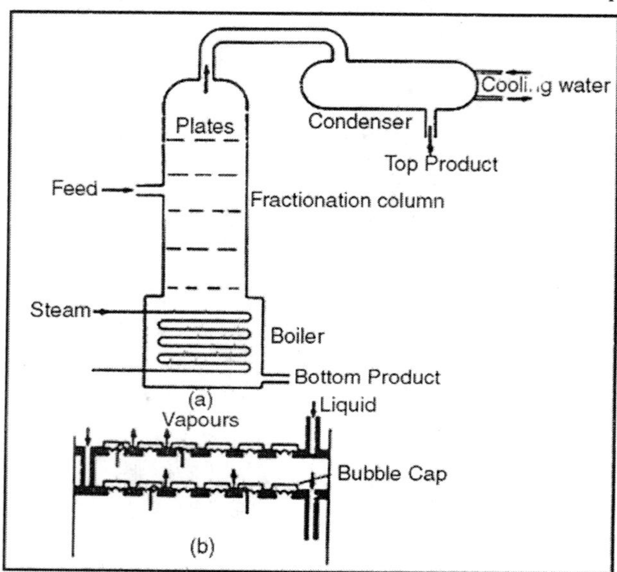

Fig. Distillation Column (a) Assembly, (b) Bubble-cap Trays

The condenser and the boiler are straightforward. The fractionation column is more complicated as it has to provide a series of contact stages for contacting the liquid and the vapour. The conventional arrangement is in the form of "bubble-cap" trays. The vapours rise through the bubble caps. The liquid flows across the trays past the bubble caps where it contacts the vapour and then over a weir and down to the next tray. Each tray represents a contact stage, or approximates to one as full equilibrium is not necessarily attained, and a sufficient number of stages must be provided to reach the desired separation of the components.

In steam distillation, the steam is bubbled through the liquid and the vapours containing the volatile component and the steam are passed to the condenser. Heat may be provided by the condensation of the steam, or independently. In some cases the steam and the condensed volatile component are immiscible, so that separation in the condenser is simple.

QUALITY CONTROL IN FOOD PROCESSING

The maintenance of a well functioning Quality Assurance (QA) programme is essential if a consistent product is to result which meets all required standards. Such a programme should be based on Hazard Analysis and Quality Analysis Critical Control Point (HACCP and QACCP) systems. HACCP and QACCP are more proactive than traditional approaches to QA/QC activities. The establishment of such programmes is the responsibility of QA personnel but the execution of it involves everyone in the company. To avoid ambiguity regarding responsibility for any QA function, it is important to assign specific HACCP/QACCP accountabilities to responsible persons and groups. A QA programme must consider all activities impacting upon product quality, from raw materials and ingredients used to product handling through distribution channels all the way to the final consumer. In respect of this, Wilson has outlined the following required components of a QA system:

- Raw material control – standard specifications must be adopted for all ingredients which must then be inspected to ensure conformity;
- Process control – all chemical, physical and microbiological hazards as well as quality factors must be identified, critical control points (CCP) must be established, monitored and a record made of any action taken;
- Finished product control–this requires that the finished product be unadulterated, properly labelled and that the integrity of the finished be protected from the environment.

HACCP AND QACCP

All food production activities must be monitored and controlled within the framework of an effective QA programme. The addition of nutrients to a food for the purpose of fortification adds to the control points which have to be considered. Poor manufacturing control leading to excessively high levels of nutrients in the finished product could have important health implications for the consumer if intake of the nutrient reaches the toxic dose.

Conversely, low levels of nutrients in the finished product could render it nutritionally ineffective. This could also have serious health implications if the target population in the fortification programme was at high nutritional risk. Poor manufacturing control could also lead to other quality defects related to interactions of added nutrients with other components of the system.

The following steps in the implementation of a quality assurance programme in the production of a fortified food have been outlined by Wilson:

- Product specifications–All specifications for fortificants, food vehicle and any other ingredients must be documented as well as acceptable deviations of these. These include specification of particle size, colour, potency, level of fortification as well as any other requirement which might be deemed necessary.
- Product safety assessment–This involves an assessment of microbiological, chemical and physical hazards for all ingredients and the finished product
- Product analysis–Sampling and testing procedures for all ingredients and the finished product must be explicitly stated.
- Determination of critical and quality control points – Based on first hand knowledge of the total process (including the plant facility, equipment and environment) stages at which inadequate control could lead to unacceptable health risk or adversely affect product quality are identified. The system of controls and actions to be taken at each control point are documented.
- Recall system–A mechanism must be put in place whereby product can be recalled if such action becomes necessary.
- QA audit–Periodic checks are necessary to verify that the QA system is effective and product quality is maintained up to the ultimate consumer.
- Feedback mechanism–Response to consumers and other relevant groups to correct any deficiencies discovered.
- Documentation of QA system–Details of the QA programme used in the production of the fortified food must be readily available to relevant individuals and organizations.

Shortcomings of many fortification programmes in the past have been due to failure to establish an adequate quality assurance programme. Evaluation of the fortification of sugar with vitamin A in Guatemala showed that only 30 per cent of samples tested were fortified at levels within the legal limits. A study of iodine content in iodised salt samples obtained from several plants in India also provided an example of the need for greater control in processing. In the determination of critical and other control points for any process accurate flow diagrams outlining the total process have been used. The construction of an accurate flow diagram for any given process requires first hand knowledge of the processing facility and its environs so that all factors which might be expected to impact on product safety could be identified.

Recommendations of the FAO/WHO Expert Technical Consultation on "The Use of HACCP in Food Control' included the following:

- Use of HACCP serves to improve food safety control and should be applied on that basis;
- The elaboration of food safety policies by government and international agencies should use risk analysis as the basis for establishing food safety priorities and for focusing inspection resources. These policies should be implemented through national strategic plans;
- In the post Uruguay Round of GATT, the Codex Alimentarius Commission should recognise the importance of its role in harmonising and establishing food standards, guidelines and recommendations particularly as it relates to safety of food in international trade. Codex should develop a strategic plan which will include a strengthening of the scientific basis for risk analysis, equivalence and the elaboration of its standards, guidelines and other recommendations and should include specific instructions to the Codex Committees on incorporating HACCP.

ANALYSIS OF VITAMINS AND MINERALS

Analysis of potency of fortificants and of vitamin and mineral content constitute an important component of the overall analytical requirements in QA/QC programmes for fortification processes. Development or selection of appropriate analytical methodologies must be based on consideration of accuracy and precision of measurements, available facilities and equipment, simplicity of procedure and rapidity of determination.

There are also many other experimental methods which have been developed in various laboratories. The Codex Committee on Methods of Analysis and Sampling is working in close cooperation with ISO, AOAC International and IUPAC to recommend protocols for determining the reliability of analytical test methods and results of laboratory analyses. The community bureau of reference of the Commission of European Communities (BCR) has set up a working group to compare analytical methods used in different laboratories in Europe.

Walter summarised the methods most commonly applied to the analysis of vitamins in foods. HPLC is the technique of choice for many of the vitamins because of the reliability of the method, the rapidity of the determination and the often reduced requirement for rigorous preliminary clean up steps. The drawbacks associated with this technology, however, are high equipment and materials costs and its lack of mobility. A comprehensive review of current methods used in the analysis of vitamins was provided by Lumely.

For analysis of vitamin A content in the MSG fortification programme Muhilal et al. developed a quantitative spectrophotometric method. A semi-quantitative method was also described suitable for field testing in the MSG-vitamin A programme. A rapid quantitative method for the determination of vitamin A content in fortified sugar has been described by Aguilar et al. This is based on the Carr-Price procedure. Adaptations of the method, yielding semi-quantitative results, suitable for field testing have also been reported.

The determination of iodine content in iodised salt has traditionally been carried out using a titrimetric method. Recently a number of HPLC based methods have been used to a large extent. One such method described by de Kleijn was used by the Food Inspection Service in the Netherlands. It utilised reversed phase HPLC with uv detection, and had a detection limit of 2.9 ng KI. Modified HPLC procedures involving precolumn derivatisation have been developed to increase the sensitivity of the analysis.

In this way Verma et al. reported a detection limit of 0.5 ng iodide. Another recent report involved a differential pulse polarographic method which did not require a separation or preconcentration step. For qualitative testing in the field, simple test kits based on the reaction of starch with iodine are available. There are important restrictions to the use of the kits which are currently being used. These are reported to be applicable over a restricted pH range.

Additives commonly used in salt such free-flowing agents or stabilisers of added iodine cause the pH to be elevated above the valid range. In such cases, false negative results are obtained. Modification of this procedure to extend the functional range of testing conditions must be investigated. Users of such field kits need to be educated as to their correct use and their limitations.

In iron fortification programmes, quite often measurement of iron content is inadequate. Bioavailability of the nutrients has been determined by measurement of labelled isotopes absorbed from the diet of human subjects or less complex methods based on animal studies such as the haemoglobin repletion method. Simple chemical methods have been used to approximate bioavailability.

There may be analyses other than the measurement of nutrient content or availability necessitated by a fortification programme. These might include the measurement of colour, particle size or moisture content as well as all testing required for the unfortified food. Based on the identification of critical and other control points, the analytical requirements of the overall quality assurance programme can be determined.

MECHANICAL SEPARATIONS

Mechanical separations can be divided into four groups—sedimentation, centrifugal separation, filtration and sieving. In sedimentation, two immiscible liquids, or a liquid and a solid, differing in density, are separated by allowing them to come to equilibrium under the action of gravity, the heavier material falling with respect to the lighter. This may be a slow process.

It is often speeded up by applying centrifugal forces to increase the rate of sedimentation; this is called centrifugal separation. Filtration is the separation of solids from liquids, by causing the mixture to flow through fine pores which are small enough to stop the solid particles but large enough to allow the liquid to pass.

Sieving, interposing a barrier through which the larger elements cannot pass, is often used for classification of solid particles.

Mechanical separation of particles from a fluid uses forces acting on these particles. The forces can be direct restraining forces such as in sieving and filtration, or indirect as in impingement filters. They can come from gravitational or centrifugal action, which can be thought of as negative restraining forces, moving the particles relative to the containing fluid. So the separating action depends on the character of the particle being separated and the forces on the particle which cause the separation. The important characteristics of the particles are size, shape and density; and of the fluid are viscosity and density.

The reactions of the different components to the forces set up relative motion between the fluid and the particles, and between particles of different character. Under these relative motions, particles and fluid accumulate in different regions and can be gathered as: in the filter cake and the filtrate tank in the filter press; in the discharge valve in the base of the cyclone and the air outlet at the top; in the outlet streams of a centrifuge; on the various sized sieves of a sieve set.

In the mechanical separations studied, the forces considered are gravity, combinations of gravity with other forces, centrifugal forces, pressure forces in which the fluid is forced away from the particles, and finally, total restraint of solid particles where normally the fluid is of little consequence. The velocities of particles moving in a fluid are important for several of these separations.

Sedimentation

Sedimentation uses gravitational forces to separate particulate material from fluid streams. The particles are usually solid, but they can be small liquid droplets, and the fluid can be either a liquid or a gas. Sedimentation is very often used in the food industry for separating dirt and debris from incoming raw material, crystals from their mother liquor and dust or product particles from air streams.

Gravitational Sedimentation of Particles in a Liquid

Solids will settle in a liquid whose density is less than their own. At low concentration, Stokes' Law will apply but in many practical instances the concentrations are too high. In a cylinder in which a uniform suspension is allowed to settle, various quite well-defined zones appear as the settling proceeds. At the top is a zone of clear liquid. Below this is a zone of more or less constant composition, constant because of the uniform settling velocity of all sizes of particles.

At the bottom of the cylinder is a zone of sediment, with the larger particles lower down. If the size range of the particles is wide, the zone of constant composition near the top will not occur and an extended zone of variable composition will replace it.

In a continuous thickener, with settling proceeding as the material flows through, and in which clarified liquid is being taken from the top and sludge from the bottom, these same zones occur. The minimum area necessary for a continuous thickener can be calculated by equating the rate of sedimentation in a particular zone to the counter-flow velocity of the rising fluid. Sedimentation equipment for separation of solid particles from liquids by gravitational sedimentation is designed to provide sufficient time for the sedimentation to occur and to permit the overflow and the sediment to be removed without disturbing the separation. Continuous flow through the equipment is generally desired, so the flow velocities have to be low enough to avoid disturbing the sediment. Various shaped vessels are used, with a sufficient cross-section to keep the velocities down and fitted with slow-speed scraper-conveyors and pumps to remove the settled solids. When vertical cylindrical tanks are used, the scrapers generally rotate about an axis in the centre of the tank and the overflow may be over a weir round the periphery of the tank.

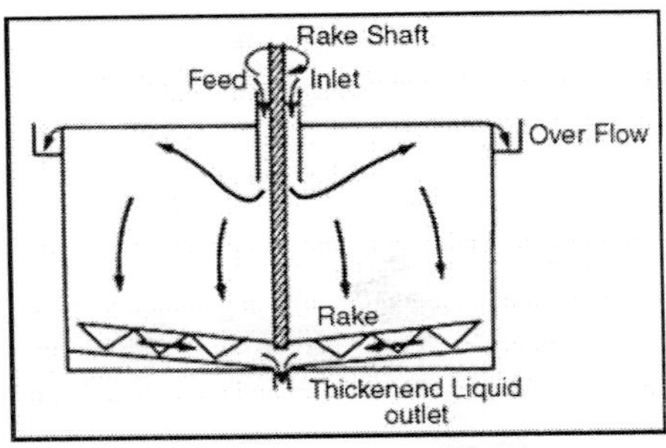

Fig. Continuous Sedimentation Plant

Flotation

In some cases, where it is not practicable to settle out fine particles, these can sometimes be floated to the surface by the use of air bubbles. This technique is known as flotation and it depends upon the relative tendency of air and water to adhere to the particle surface. The water at the particle surface must be displaced by air, after which the buoyancy of the air is sufficient to carry both the particle and the air bubble up through the liquid.

Because it depends for its action upon surface forces, and surface forces can be greatly changed by the presence of even minute traces of surface active agents, flotation may be promoted by the use of suitable additives. In some instances, the air bubbles remain round the solid particles and cause froths. These are produced in vessels fitted with mechanical agitators, the agitators whip up the air-liquid mixture and overflow the froth into collecting troughs.

The greatest application of froth flotation is in the concentration of minerals, but one use in the food industry is in the separation of small particles of fat from water. Dissolving the air in water under pressure provides the froth. On the pressure being suddenly released, the air comes out of solution in the form of fine bubbles which rise and carry the fat with them to surface scrapers.

Sedimentation of Particles in a Gas

An important application, in the food industry, of sedimentation of solid particles occurs in spray dryers. In a spray dryer, the material to be dried is broken up into small droplets of about 100 mm diametre and these fall through heated air, drying as they do so. The necessary area so that the particles will settle can be calculated in the same way as for sedimentation. Two disadvantages arise from the slow rates of sedimentation: the large chamber areas required and the long contact times between particles and the heated air which may lead to deterioration of heat-sensitive products.

Settling under Combined Forces

It is sometimes convenient to combine more than one force to effect a mechanical separation. In consequence of the low velocities, especially of very small particles, obtained when gravity is the only external force acting on the system, it is well worthwhile to also employ centrifugal forces. Probably the most common application of this is the cyclone separator. Combined forces are also used in some powder classifiers such as the rotary mechanical classifier and in ring dryers.

Impingement Separators

Other mechanical flow separators for particles in a gas use the principal of impingement in which deflector plates or rods, normal to the direction of flow of the stream, abruptly change the direction of flow.

The gas recovers its direction of motion more rapidly than the particles because of its lower inertia. Suitably placed collectors can then be arranged to collect the particles as they are thrown out of the stream.

This is the principle of operation of mesh and fibrous air filters. Various adaptations of impingement and settling separators can be adapted to remove particles from gases, but where the particle diameters fall below about 5 mm, cloth filters and packed tubular filters are about the only satisfactory equipment.

Classifiers

Classification implies the sorting of particulate material into size ranges. Use can be made of the different rates of movement of particles of different sizes and densities suspended in a fluid and differentially affected by imposed forces such as gravity and centrifugal fields, by making suitable arrangements to collect the different fractions as they move to different regions.

Rotary mechanical classifiers, combining differential settling with centrifugal action to augment the force of gravity and to channel the size fractions so that they can be collected, have come into increasing use in flour milling. One result of this is that because of small differences in sizes, shapes and densities between starch and protein-rich material after crushing, the flour can be classified into protein-rich and starch-rich fractions. Rotary mechanical classifiers can be used for other large particle separation in gases.

Classification is also employed in direct air dryers, in which use is made of the density decrease of material on drying. Dry material can be sorted out as a product and wet material returned for further drying. One such dryer uses a scroll casing through which the mixed material is passed, the wet particles pass to the outside of the casing and are recycled while the material in the centre is removed as dry product.

CENTRIFUGAL SEPARATIONS

The separation by sedimentation of two immiscible liquids, or of a liquid and a solid, depends on the effects of gravity on the components. Sometimes this separation may be very slow because the specific gravities of the components may not be very different, or because of forces holding the components in association, for example, as occur in emulsions.

Also, under circumstances when sedimentation does occur there may not be a clear demarcation between the components but rather a merging of the layers.

For example, if whole milk is allowed to stand, the cream will rise to the top and there is eventually a clean separation between the cream and the skim milk. However, this takes a long time, of the order of one day, and so it is suitable, perhaps, for the farm kitchen but not for the factory. Much greater forces can be obtained by introducing centrifugal action, in a centrifuge.

Gravity still acts and the net force is a combination of the centrifugal force with gravity as in the cyclone. Because in most industrial centrifuges, the centrifugal forces imposed are so much greater than gravity, the effects of gravity can usually be neglected in the analysis of the separation.

The centrifugal force depends upon the radius and speed of rotation and upon the mass of the particle. If the radius and the speed of rotation are fixed, then the controlling factor is the weight of the particle so that the heavier the particle the greater is the centrifugal force acting on it. Consequently, if two liquids, one of which is twice as dense as the other, are placed in a bowl and the bowl is rotated about a vertical axis at high speed, the centrifugal force per unit volume will be twice as great for the heavier liquid as for the lighter.

The heavy liquid will therefore move to occupy the annulus at the periphery of the bowl and it will displace the lighter liquid towards the centre. This is the principle of the centrifugal liquid separator.

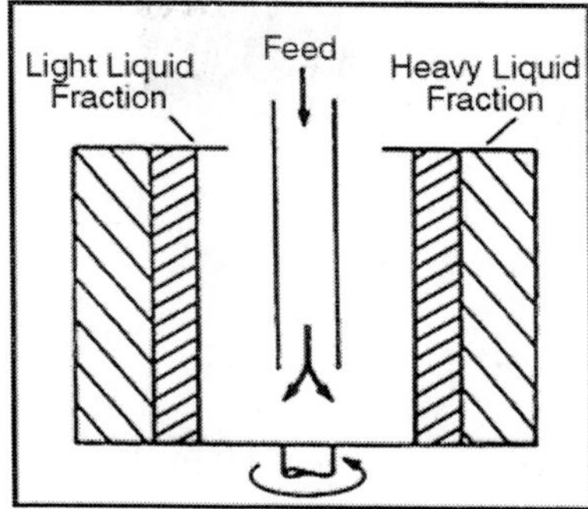

Fig. Liquid Separation in a Centrifuge

Liquid Separation

The separation of one component of a liquid-liquid mixture, where the liquids are immiscible but finely dispersed, as in an emulsion, is a common operation in the food industry. It is particularly common in the dairy industry in which the emulsion, milk, is separated by a centrifuge into skim milk and cream. It seems worthwhile, on this account, to examine the position of the two phases in the centrifuge as it operates. The milk is fed continuously into the machine, which is generally a bowl rotating about a vertical axis, and cream and skim milk come from the respective discharges. At some point within the bowl there must be a surface of separation between cream and the skim milk.

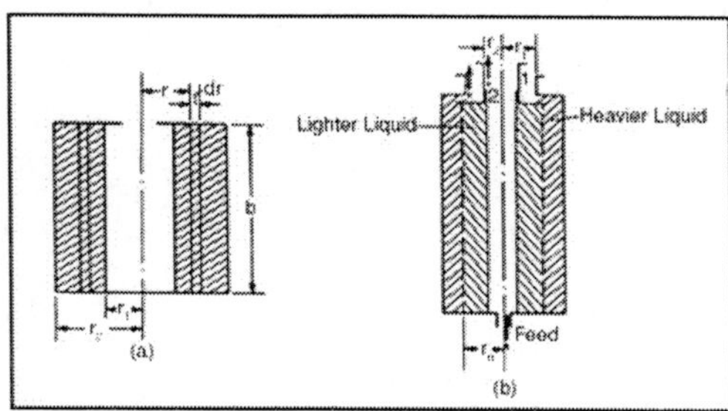

Fig. Liquid Centrifuge (a) Pressure Difference (b) Neutral Zone

Filtration

In another class of mechanical separations, placing a screen in the flow through which they cannot pass imposes virtually total restraint on the particles above a given size. The fluid in this case is subject to a force that moves it past the retained particles. This is called filtration. The particles suspended in the fluid, which will not pass through the apertures, are retained and build up into what is called a filter cake. Sometimes it is the fluid, the filtrate, that is the product, in other cases the filter cake.

The fine apertures necessary for filtration are provided by fabric filter cloths, by meshes and screens of plastics or metals, or by beds of solid particles. In some cases, a thin preliminary coat of cake, or of other fine particles, is put on the cloth prior to the main filtration process.

This preliminary coating is put on in order to have sufficiently fine pores on the filter and it is known as a pre-coat. The analysis of filtration is largely a question of studying the flow system. The fluid passes through the filter medium, which offers resistance to its passage, under the influence of a force which is the pressure differential across the filter.

Thus, we can write the familiar equation:

Rate of filtration = Driving force/resistance

Resistance arises from the filter cloth, mesh, or bed, and to this is added the resistance of the filter cake as it accumulates. The filter-cake resistance is obtained by multiplying the specific resistance of the filter cake, that is its resistance per unit thickness, by the thickness of the cake. The resistances of the filter material and pre-coat are combined into a single resistance called the filter resistance. It is convenient to express the filter resistance in terms of a fictitious thickness of filter-cake. This thickness is multiplied by the specific resistance of the filter-cake to give the filter resistance.

Filtration Equipment

The basic requirements for filtration equipment are:

- Mechanical support for the filter medium.
- Flow accesses to and from the filter medium.
- Provision for removing excess filter-cake.

In some instances, washing of the filter-cake to remove traces of the solution may be necessary. Pressure can be provided on the upstream side of the filter, or a vacuum can be drawn downstream, or both can be used to drive the wash fluid through.

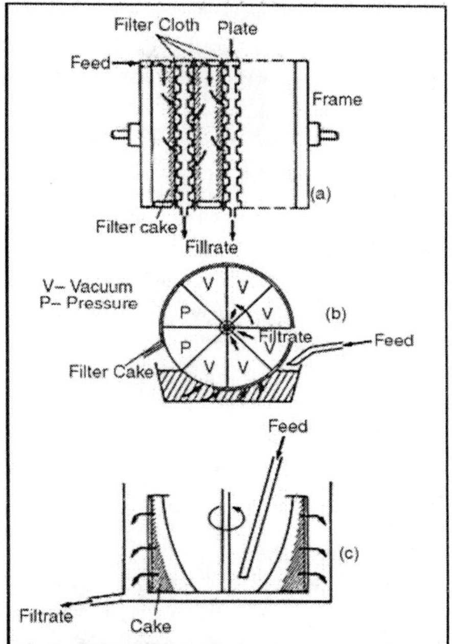

Fig. Filtration Equipment: (a) Plate and Frame Press (b) Rotary Vacuum Filter (c) Centrifugal Filter

Plate and Frame Filter Press

In the plate and frame filter press, a cloth or mesh is spread out over plates which support the cloth along ridges but at the same time leave a free area, as large as possible, below the cloth for flow of the filtrate. The plates with their filter cloths may be horizontal, but they are more usually hung vertically with a number of plates operated in parallel to give sufficient area.

Filter cake builds up on the upstream side of the cloth that is the side away from the plate. In the early stages of the filtration cycle, the pressure drop across the cloth is small and filtration proceeds at more or less a constant rate. As the cake increases, the process becomes more and more a constant-pressure one and this is the case throughout most of the cycle. When the available space between successive frames is filled with cake, the press has to be dismantled and the cake scraped off and cleaned, after which a further cycle can be initiated.

The plate and frame filter press is cheap but it is difficult to mechanize to any great extent. Variants of the plate and frame press have been developed which allow easier discharging of the filter-cake. For example, the plates, which may be rectangular or circular, are supported on a central hollow shaft for the filtrate and the whole assembly enclosed in a pressure tank containing the slurry.

Filtration can be done under pressure or vacuum. The advantage of vacuum filtration is that the pressure drop can be maintained whilst the cake is still under atmospheric pressure and so can be removed easily. The disadvantages are the greater costs of maintaining a given pressure drop by applying a vacuum and the limitation on the vacuum to about 80 kPa maximum. In pressure filtration, the pressure driving force is limited only by the economics of attaining the pressure and by the mechanical strength of the equipment.

CONCENTRATIONS

The driving force, which produces equilibrium distributions, is considered to be proportional at any time to the difference between the actual concentration and equilibrium concentration of the component being separated. Thus, concentrations in contact equilibrium separation processes are linked with the general driving force concept. Consider a case in which initially all of the molecules of some component *A* of a gas mixture are confined by a partition in one region of a system. The partition is then removed. Random movement among the gas molecules will, in time, distribute component *A* through the mixture. The greater the concentration of *A* in the partitioned region, the more rapidly will diffusion occur across the boundary once the partition is removed. The relative proportions of the components in a mixture or a solution are expressed in terms of the concentrations. Any convenient units may be used for concentration, such as g g^{-1}, g kg^{-1}, mg g^{-1}, percentages, parts per million, and so on.

Because the gas laws are based on numbers of molecules, it is often convenient to express concentrations in terms of the relative numbers of molecules of the components. The mole fraction of a component in a mixture is the proportion of the number of molecules of the component present to the total number of the molecules of all the components.

Gas/Liquid Equilibria

Molecules of the components in a liquid mixture or solution have a tendency to escape from the liquid surface into the gas above the solution. The escaping tendency sets up a pressure above the surface of the liquid owing to the resultant concentration of the escaped molecules. This pressure is called the vapour pressure of the liquid. The magnitude of the vapour pressure depends upon the liquid composition and upon the temperature. For a solution or a mixture, the various components in the liquid each exert their own partial vapour pressures. When the liquid contains various components it has been found that, in many cases, the partial vapour pressure of any component is proportional to the mole fraction of that component in the liquid.

Solid/Liquid Equilibra

Liquids have a capacity to dissolve solids up to an extent, which is determined by the solubility of the particular solid material in that liquid. Solubility is a function of temperature and, in most cases, solubility increases with rising temperature. A solubility curve can be drawn to show this relationship, based on the equilibrium concentration in solution measured in convenient units, for example gkg^{-1}, as a function of temperature. Such a curve is illustrated in Fig. for sodium nitrite in water. There are some relatively rare systems in which solubility decreases with temperature, and they provide what is termed a reversed solubility curve. The equilibrium solution, which is ultimately reached between solute and solvent, is called a saturated solution, implying that no further solute can be taken into solution at that particular temperature.

An unsaturated solution plus solid solute is not in equilibrium, as the solvent can dissolve more of the solid. When a saturated solution is heated, if it has a normal solubility curve the solution then has a capacity to take up further solute material, and so it becomes unsaturated.

Conversely, when a saturated solution is cooled it becomes super saturated, and at equilibrium that solute which is in excess of the solubility level at the particular temperature will come out of solution, normally as crystals.

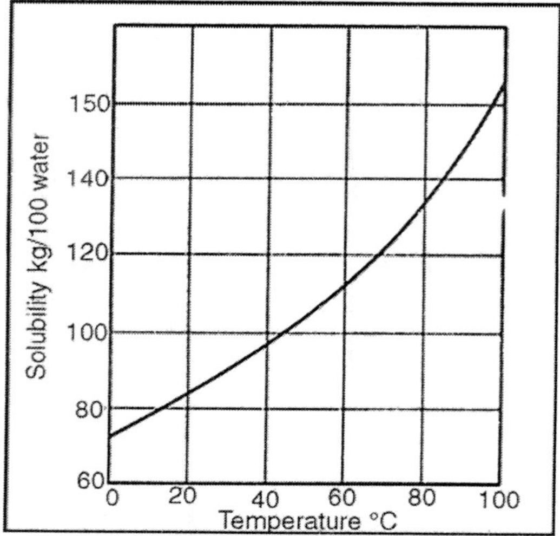

Fig. Solubility of Sodium Nitrite in Water

However, this may not occur quickly and in the interim the solution is spoken of as super saturated and it is said to be in a metastable state.

OVERALL VIEW OF AN ENGINEERING PROCESS

Using a material balance and an energy balance, a food engineering process can be viewed overall or as a series of units. Each unit is a unit operation. The unit operation can be represented by a box as shown in Figure.

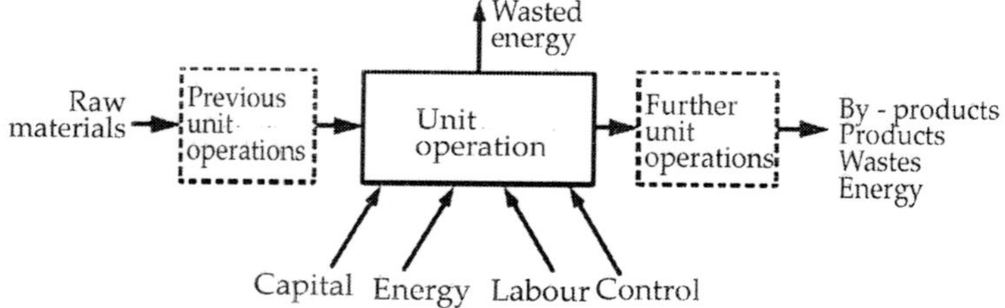

Fig. Unit Operation

Into the box go the raw materials and energy, out of the box come the desired products, by-products, wastes and energy. The equipment within the box will enable the required changes to be made with as little waste of materials and energy as possible. In other words, the desired products are required to be maximized and the undesired by-products and wastes minimized. Control over the process is exercised by regulating the flow of energy, or of materials, or of both.

EQUILIBRIUM-CONCENTRATION RELATIONSHIPS

A contact equilibrium-separation process is designed to reduce the concentration of a component in one phase, or flowing stream in a continuing process, and increase it in another. Conventionally, and just to distinguish one stream from another, one is called the overflow and the other the underflow. The terms referred originally to a system of two immiscible liquids, one lighter (the overflow) and the other heavier (the underflow) than the other, and between which the particular component was transferred. When there are several stages, the overflow and underflow streams then move off in opposite directions in a counter flow system.

Following standard chemical engineering nomenclature, the concentration of the component of interest in the lighter stream, that is the stream with the lower density, is denoted by y. For example, in a gas absorption system, the light stream would be the gas; in a distillation column, it would be the vapour stream; in a liquid-extraction system, it would be the liquid overflow. The concentration of the component in the heavier stream is denoted by x. Thus we have two streams; in one the concentration of the component is y and in the other, the heavier stream, it is x.

In the simple case of multistage oil extraction with a solvent, equilibrium is generally attained in each stage. The concentration of the oil is the same in the liquid solution spilling over or draining off in the overflow as it is in the liquid in the underflow containing the solids, so that in this case $y = x$ and the equilibrium/concentration diagram is a straight line. In gas absorption, such relationships as Henry's Law relate the concentration in the light gas phase to that in the heavy liquid phase and so enable the equilibrium diagram to be plotted.

In crystallization, the equilibrium concentration corresponds to the solubility of the solute at the particular temperature. Across a membrane there is some equilibrium distribution of the particular component of interest. If the concentration in one stream is known, the equilibrium diagram allows us to read off the corresponding concentration in the other stream if equilibrium has been attained.

The attainment of equilibrium takes time and this has to be taken into account when considering contact stages. The usual type of rate equation applies, in which the rate is given by the driving force divided by a resistance term. The driving force is the extent of the departure from equilibrium and generally is measured by concentration differences. Resistances have been classified in different ways but they are generally assumed to be concentrated at the phase boundary.

In many practical cases, allowance can be made for non-attained equilibrium by assuming an efficiency for each stage, in effect a percentage of equilibrium actually attained.

Gas Absorption

Gas absorption/desorption is a process in which a gaseous mixture is brought into contact with a liquid and during this contact a component is transferred between the gas stream and the liquid stream.

The gas may be bubbled through the liquid, or it may pass over streams of the liquid, arranged to provide a large surface through which the mass transfer can occur. The liquid film in this latter case can flow down the sides of columns or over packing, or it can cascade from one tray to another with the liquid falling and the gas rising in the counter flow.

The gas, or components of it, either dissolves in the liquid (absorption) or extracts a volatile component from the liquid (desorption). An example of the first type is found in hydrogenation of oils, in which the hydrogen gas is bubbled through the oil with which it reacts.

Generally, there is a catalyst present also to promote the reaction. The hydrogen is absorbed into the oil, reacting with the unsaturated bonds in the oil to harden it. Another example of gas absorption is in the carbonation of beverages. Carbon dioxide under pressure is dissolved in the liquid beverage, so that when the pressure is subsequently released on opening the container, effervescence occurs.

An example of desorption is found in the steam stripping of fats and oils in which steam is brought into contact with the liquid fat or oil, and undesired components of the fat or oil pass out with the steam.

This is used in the deodorizing of natural oils before blending them into food products such as margarine, and in the stripping of unwanted flavours from cream before it is made into butter. The equilibrium conditions arise from the balance of concentrations of the gas or the volatile flavour, between the gas and the liquid streams.

In the gas absorption process, sufficient time must be allowed for equilibrium to be attained so that the greatest possible transfer can occur and, also, opportunity must be provided for contacts between the streams to occur under favourable conditions.

Stage Equilibrium Gas Absorption

The performance of counter current stage contact gas absorption equipment can be calculated if the operating and equilibrium conditions are known. The liquid stream and the gas stream are brought into contact in each stage and it is assumed that sufficient time is allowed for equilibrium to be reached. In cases where sufficient time is not available for equilibration, the rate equations have to be introduced and this complicates the analysis.

However, in many cases of practical importance in the food industry, either the time is sufficient to reach equilibrium, or else the calculation can be carried out on the assumption that it is and a stage efficiency term, a fractional attainment of equilibrium, introduced to allow for the conditions actually attained. Appropriate efficiency values can sometimes be found from published information, or sought experimentally.

After the streams in a contact stage have come to equilibrium, they are separated and then pass in opposite directions to the adjacent stages. The separation of the gas and the liquid does not generally present great difficulty and some form of cyclone separator is often used.

In order to calculate the equipment performance, operating conditions must be known or found from the mass balances.

Very often the known factors are:

- Gas and liquid rates of flow.
- Inlet conditions of gas and liquid.
- One of the outlet conditions.
- Equilibrium relationships.

Gas Absorption Equipment

Gas absorption equipment is designed to achieve the greatest practicable interfacial area between the gas and the liquid streams, so that liquid sprays and gas-bubbling devices are often employed. In many cases, a vertical array of trays is so arranged that the liquid descends over a series of perforated trays, or flows down over ceramic packing that fills a tower.

For the hydrogenation of oils, absorption is followed by reaction of the hydrogen with the oil, and a nickel catalyst is used to speed up the reactions. Also, pressure is applied to increase gas concentrations and therefore speed up the reaction rates. Practical problems are concerned with arranging distribution of the catalyst, as well as of oil and hydrogen. Some designs spray oil and catalyst into hydrogen, others bubble hydrogen through a continuous oil phase in which catalyst particles are suspended. For the stripping of volatile flavours and taints in deodorizing equipment, the steam phase is in general the continuous one and the liquid is sprayed into this and then separated. In one design of cream deodorizing plant, cream is sprayed into an atmosphere of steam and the two streams then pass on to the next stages, or the steam may be condensed and fresh steam used in the next stage.

Extraction and Washing

It is often convenient to use a liquid in order to carry out a separation process. The liquid is thoroughly mixed with the solids or other liquid from which the component is to be removed and then the two streams are separated.

In the case of solids, the separation of the two streams is generally by simple gravity settling. Sometimes it is the solution in the introduced liquid that is the product required, such as in the extraction of coffee from coffee beans with water. In other cases, the washed solid may be the product as in the washing of butter. The term washing is generally used where an unwanted constituent is removed in a stream of water. Extraction is also an essential stage in the sugar industry when soluble sucrose is removed by water extraction from sugarcane or beet. Washing occurs so frequently as to need no specific examples. To separate liquid streams, the liquids must be immiscible, such as oil and water. Liquid-liquid extraction is the name used when both streams in the extraction are liquid. Examples of extraction are found in the edible oil industry in which oil is extracted from natural products such as peanuts, soya beans, rapeseeds, sunflower seeds and so on. Liquid-liquid extraction is used in the extraction of fatty acids.

Washing

Washing is almost identical to extraction, the main distinction being one of the emphasis in that in washing the inert material is the required product, and the solvent used is water which is cheap and readily available. Various washing situations are encountered and can be analysed. That to be considered is one in which a solid precipitate, the product, retains water which also contains residues of the mother liquor so that on drying without washing these residues will remain with the product.

The washing is designed to remove them, and examples are butter and casein and cheese washing in the dairy industry. Calculations on counter current washing can be carried out using the same methods as discussed under extraction, working from the operating and equilibrium conditions. In washing, fresh water is often used for each stage and the calculations for this are also straightforward.

Extraction and Washing Equipment

The first stage in an extraction process is generally mechanical grinding, in which the raw material is shredded, ground or pressed into suitably small pieces or flakes to give a large contact area for the extraction. In some instances, for example in sugarcane processing and in the extraction of vegetable oils, a substantial proportion of the desired products can be removed directly by expression at this stage and then the remaining solids are passed to the extraction plant. Fluid solvents are easy to pump and so overflows are often easier to handle than underflows and sometimes the solids may be left and solvent from successive stages brought to them.

This is the case in the conventional extraction battery. In this a number of tanks, each suitable both for mixing and for settling, are arranged in a row or a ring. The solids remain in the one mixer-settler and the solvent is moved progressively round the ring of tanks, the number, n, often being about 12. At any time, two of the tanks are out of operation, one being emptied and the other being filled.

In the remaining (n - 2), tanks extraction is proceeding with the extracting liquid solvent, usually water, being passed through the tanks in sequence, the "oldest" (most highly extracted) tank receiving the fresh liquid and the "youngest" (newly filled with fresh raw material) tank receiving the most concentrated liquid. After leaving this "youngest" tank, the concentrated liquid passes from the extraction battery to the next stage of the process. After a suitable interval, the connections are altered so that the tank which has just been filled becomes the new "youngest" tank.

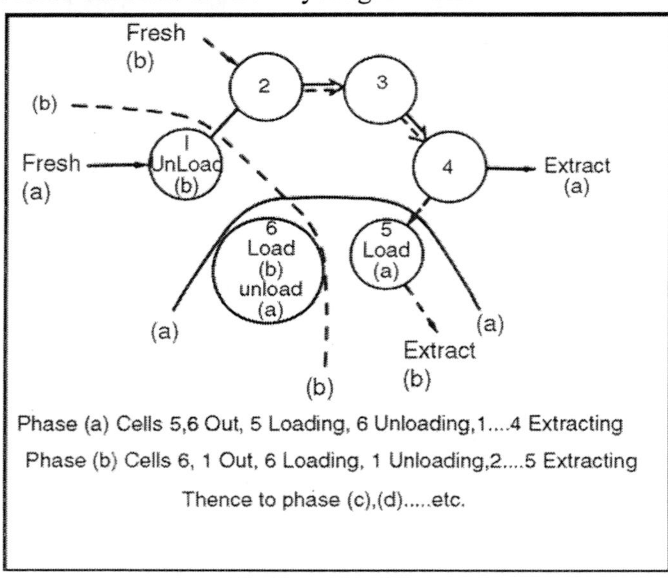

Fig. Extraction Battery

The former "oldest" tank comes out of the sequence and is emptied, the one that was being emptied is filled and the remaining tanks retain their sequence but with each becoming one stage "older". This procedure which is illustrated in Fig in effect accomplishes counter current extraction, but with only the liquid physically having to be moved apart from the emptying and filling in the terminal tanks.

In the same way and for the same reasons as with counter flow heat exchangers, this counter current (or counter flow) extraction system provides the maximum mean driving force, the log mean concentration difference in this case, contrasting with the log mean temperature difference in the heat exchanger. This ensures that the equipment is used efficiently.

In some other extractors, the solids are placed in a vertical bucket conveyor and moved up through a tower down which a stream of solvent flows. Other forms of conveyor may also be used, such as screws or metal bands, to move the solids against the solvent flow. Sometimes centrifugal forces are used for conveying, or for separating after contacting.

Washing is generally carried out in equipment that allows flushing of fresh water over the material to be washed. In some cases, the washing is carried out in a series of stages. Although water is cheap, in many cases very large quantities are used for washing so that attention paid to more efficient washing methods may well be worthwhile. Much mechanical ingenuity has been expended upon equipment for washing and many types of washers are described in the literature.

CRYSTALLIZATION

Crystallization is an example of a separation process in which mass is transferred from a liquid solution, whose composition is generally mixed, to a pure solid crystal. Soluble components are removed from solution by adjusting the conditions so that the solution becomes supersaturated and excess solute crystallizes out in a pure form. This is generally accomplished by lowering the temperature, or by concentration of the solution, in each case to form a supersaturated solution from which crystallization can occur. The equilibrium is established between the crystals and the surrounding solution, the mother liquor.

The manufacture of sucrose, from sugarcane or sugar beet, is an important example of crystallization in food technology. Crystallization is also used in the manufacture of other sugars, such as glucose and lactose, in the manufacture of food additives, such as salt, and in the processing of foodstuffs, such as ice cream. In the manufacture of sucrose from cane, water is added and the sugar is pressed out from the residual cane as a solution. This solution is purified and then concentrated to allow the sucrose to crystallize out from the solution.

Crystallization Equilibrium

Once crystallization is concluded, equilibrium is set up between the crystals of pure solute and the residual mother liquor, the balance being determined by the solubility (concentration) and the temperature. The driving force making the crystals grow is the concentration excess (supersaturation) of the solution above the equilibrium (saturation) level. The resistances to growth are the resistance to mass transfer within the solution and the energy needed at the crystal surface for incoming molecules to orient themselves to the crystal lattice.

Heat of Crystallization

When a solution is cooled to produce a supersaturated solution and hence to cause crystallization, the heat that must be removed is the sum of the sensible heat necessary to cool the solution and the heat of crystallization. When using evapouration to achieve the supersaturation, the heat of vapourization must also be taken into account. Because few heats of crystallization are available, it is usual to take the heat of crystallization as equal to the heat of solution to form a saturated solution. Theoretically, it is equal to the heat of solution plus the heat of dilution, but the latter is small and can be ignored. For most food materials, the heat of crystallization is positive, i.e., heat is given out during crystallization. Note that heat of crystallization is the opposite of heat of solution. If a material takes in heat, i.e., has a negative heat of solution, then the heat of crystallization is positive. Heat balances can be calculated for crystallization.

Stage-Equilibrium Crystallization

When the first crystals have been separated, the mother liquor can have its temperature and concentration changed to establish a new equilibrium and so a new harvest of crystals. The limit to successive crystallizations is the build up of impurities in the mother liquor which makes both crystallization and crystal separation slow and difficult.

This is also the reason why multiple crystallizations are used, with the purest and best crystals coming from the early stages. For example, in the manufacture of sugar, the conc-ntration of the solution is increased and then seed crystals are added. The temperature is controlled until the crystal nucleii added have grown to the desired size, then the crystals are separated from the residual liquor by centrifuging. The liquor is next returned to a crystallizing evapourator, concentrated again to produce further supersaturation, seeded and a further crop of crystals of the desired size grown. By this method the crystal size of the sugar can be controlled. The final mother liquor, called molasses, can be held indefinitely without producing any crystallization of sugar.

Crystallization Equipment

Crystallizers can be divided into two types: crystallizers and evapourators. A crystallizer may be a simple open tank or vat in which the solution loses heat to its surroundings. The solution cools slowly so that large crystals are generally produced. To increase the rate of cooling, agitation and cooling coils or jackets are introduced and these crystallizers can be made continuous. The simplest is an open horizontal trough with a spiral scraper. The trough is water jacketed so that its temperature can be controlled. An important crystallizer in the food industry is the cylindrical, scraped surface heat exchanger, which is used for plasticizing margarine and cooking fat, and for crystallizing ice cream.

It is essentially a double-pipe heat exchanger fitted with an internal scraper. The material is pumped through the central pipe and agitated by the scraper, with the cooling medium flowing through the annulus between the outer pipes.

A crystallizer in which considerable control can be exercised is the Krystal or Oslo crystallizer. In this, a saturated solution is passed in a continuous cycle through a bed of crystals. Close control of crystal size can be obtained. Evapourative crystallizers are common in the sugar and salt industries. They are generally of the calandria type. Vacuum evapourators are often used for crystallization as well, though provision needs to be made for handling the crystals. Control of crystal size can be obtained by careful manipulation of the vacuum and feed.

The evapourator first concentrates the sugar solution, and when seeding commences the vacuum is increased. This increase causes further evapouration of water which cools the solution and the crystals grow. Fresh saturated solution is added to the evapourator and evapouration continued until the crystals are of the correct size. In some cases, open pan steam-heated evapourators are still used, for example, in making coarse salt for the fish industry. In some countries, crystallization of salt from sea water is effected by solar energy which concentrates the water slowly and this generally gives large crystals.

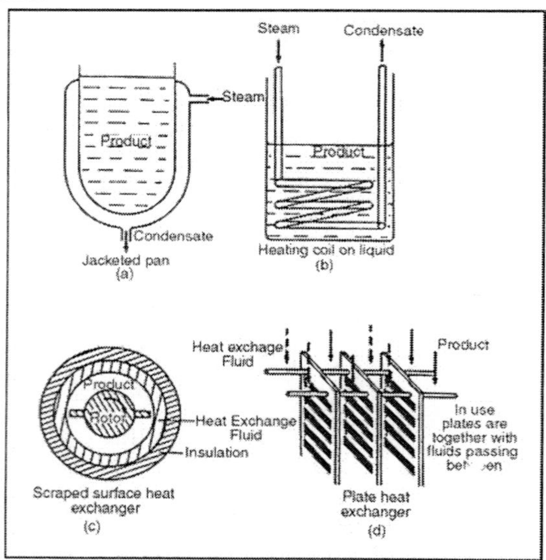

Fig. Heat Exchange Equipment

Crystals are regular in shape: cubic, rhombic, tetragonal and so on. The shape of the crystals forming may be influenced by the presence of other compounds in the solution, even in traces.

The shape of the crystal is technologically important because such properties as the angle of repose of stacked crystals and rate of dissolving are related to the crystal shape. Another important property is the uniformity of size of the crystals in a product. In a product such as sucrose, a non-uniform crystalline mixture is unattractive in appearance, and difficult to handle in packing and storing as the different sizes tend to separate out. Also the important step of separating mother liquor from the crystals is more difficult.

Chapter 7

Food Processing and Packaging

Food processing operations now being conducted or proposed include cook-chill; vacuum packaging; sous vide; smoking and curing; brewing, processing, and bottling alcoholic beverages, carbonated beverages, or drinking water; and custom processing of animals. The Food Code specifies that a HACCP plan acceptable to the regulatory authority be the basis for approving food manufacturing/processing operations at retail.

MARKET AND GROWTH DRIVERS

The global market for food processing enzymes in 2002 was approximately US $700 million that increased to US $740 million in the year 2004. Food and feed enzymes together had nearly 45 per cent share, followed by industrial enzymes which are used in the detergent industry.

In terms of value, this is contributing approx. US $450 million in the world enzyme market estimated at US $1000 million by the year 2005. However, food processing enzymes find limited applications in India and hence its market also has remained un-tapped till date. Enzymes for food and feed processing industries contribute almost 17 per cent to the Indian market. In recent years, the trend is changing with globalization of food processing industry in India, Food processing enzyme market is fast growing in diverse applications. Biocon Ltd and Novozymes are the two prominent players in the Indian food processing enzymes industry, their sales turnover for 2003-04 and 2004-05 and growth rate is summarized here below:

Table 1. Sales INR in Millions

Sr. No.	Name of Company	2003-2004	2004-05	Growth (%)
1.	Biocon Ltd.	5020.0	6464.0	29
2.	Novozymes Bangalore	530.0	690.0	30

GROWTH DRIVERS

- Population growth, urbanization and rising income levels have increased demand on food items with its spill-over effect seen on the growth of demand of food products and this in turn has spurred the demand of food enzymes in India.
- Technical innovation in enzymatic food processing is a major growth driver for this segment.
- The fast changing food habits, increased use of fast foods, modified foods and instant foods are other drivers for demand of food processing enzymes.
- Health food is a sunrise industry in India and also globally, where food processing enzymes find wider application in manufacturing of these products.

TECHNOLOGY/PROCESS

Fermentation

- Fermentation of a selected microorganism under sterile conditions.
- Production of the desired enzyme by consumption of carbohydrates, proteins, salts, water and energy
- Transfer of the fermented broth to recovery.

Recovery

- Filtration of the fermentation broth.
- Purification of the liquid phase containing the enzyme.
- Concentration of the liquid enzyme to desired enzymatic activity.
- Stabilization and standardization of a liquid product.
- Standardization of a liquid concentrate for granulation.

AGRICULTURE AND FOOD PROCESSING

SOURCES OF FINANCE FOR AGRICULTURE

At present nearly 60% of rural credit is used for agriculture and allied activities, 10% for non-farm activities and the balance 30% for funding household consumption expenditure. Nearly 40 - 45% of rural credit needs are catered to by formal credit institutions and the balance by the informal sector including commission agents, input suppliers, traders, Self Help Groups (Joint Liability Groups), processing industries and professional money lenders.

The Indian commercial banks are mandated by the Reserve Bank of India (RBI), to undertake directed lending for agriculture and rural development. Commercial banks are required to achieve priority sector lending of 40% of net bank credit of which 18% should be for agriculture. Further, sub-targets are also specified *i.e.*, Direct Agriculture (credit for direct farmer benefit, Target: Minimum of 13.5% of the net bank credit) and Indirect Agriculture (Target: Maximum of 4.5% of the net bank credit).

In FY 03, total agricultural advances outstanding under priority sector lending, by Public Sector Banks stood at INR 735 billion, on a total net bank credit portfolio of INR 4779 billion. As compared to this, agricultural advances by Private Sector Banks stood at INR 119 billion, on a total net bank credit portfolio of INR 718 billion. The following Exhibit provides details of direct and indirect priority sector credit amongst Public and Private sector banks during FY03.

In case of a shortfall in lending to agriculture, banks are required to invest in Rural Infrastructure Development Fund (RIDF) deposits, which offers interest rates linked to the bank's performance in lending to agriculture. These rates are inversely proportional to the shortfall in agricultural lending. The interest rate on RIDF is provided in slabs, based on the achievement of the priority sector lending target, and can be as low as 4% to 5%.Therefore, it is imperative to achieve the agriculture target and more specifically the Direct Agriculture target as it accounts for 75% of the total.

Established in 1996, RIDF is being utilized for making investments in rural infrastructure projects. Nine tranches of RIDF have been established with an aggregate corpus of INR 340 billion, with the ninth tranche of INR 55 billion. Cumulative sanctions and disbursements under various tranches of RIDF stood at INR 334 billion and INR 193 billion respectively, at the end of February 2004.The RIDF funds are used to finance minor irrigation projects, shallow wells, drinking water facilities, roads, bridges and storage infrastructure.

SOURCE OF FINANCE AND FOOD PROCESSING

Banks: The Food Processing sector has access to credit from Commercial Banks- Indian and Foreign, Cooperative Banks and the Regional Rural Banks. The formal institutional sector assists the food processing companies through long term loans for capital investments and short term loans for working capital. The total lending to the food processing sector is clubbed with lending to the agriculture sector and separate details for financing to the food processing sector are not available.

- National Bank for Agricultural and Rural Development (NABARD): NABARD offers refinance facilities for food processing, agri infrastructure, developmental assistance for RRBs, DCCBs, SSI and SHG linkages, and assists in infrastructure development and research.
- Small Industries Development Bank of India (SIDBI):SIDBI has been involved in assisting the entire SSI Sector including tiny, village and cottage industries through suitable schemes tailored to address the funding requirements for expansion, diversification, modernisation and rehabilitation.
- Export Import Bank (EXIM Bank): Exim Bank assists in financing and facilitation of foreign trade. The Bank also offers financial support to companies engaged in exports.
- National Cooperative Development Corporation (NCDC): NCDC assists in promotion, planning and financing the agricultural supply chain from production, processing, storage and trade of agricultural produce and food products. NCDC also provides assistance for marketing of certain notified commodities *e.g.* fertilizers, pesticides, agricultural machinery etc.
- Ministries/Government bodies
- Ministry of Food Processing Industries (MFPI): MFPI is the Nodal agency for development of the processed food sector

in the country MFPI's financial schemes include schemes for technology upgradation, Human resource development, Quality testing, R&D,TQM, backward and forward integration, development of infrastructure including food parks.

- Agricultural and Processed Foods Products Export Development Authority (APEDA): APEDA facilitates market linkages between Indian producers, manufacturers and the international market. APEDA provides financial assistance for market development, infrastructure development and development of quality enhancing facilities.
- Ministry of Agriculture (MoA), Government of India: The Ministry of Agriculture under various schemes, provides financial assistance for development of specific crops for investment in seeds, irrigation, farm implements, inputs, infrastructure and training.
- National Horticultural Board (NHB):NHB promotes integrated development in horticulture, assists in development of post harvest management infrastructure, promotes production and processing of fruits and vegetables, strengthening of market information systems and assists in R&D programmes in cultivation and processing. NHB's financial schemes are directed towards commercial horticulture and infrastructure related to post harvest techniques.

Financial assistance from these organizations are in the form of grants, back-ended subsidies, soft loans, refinance etc., with most of the schemes directed to specific sub-sectors of the agri/food processing industry. In addition, some State Governments offer financial schemes to companies in the sector.

The key issues with Government schemes are highlighted below:

- Overlap among schemes: Various Government agencies have separate schemes which aim to achieve the same objectives. Thus, an applicant can potentially approach different agencies and obtain assistance from more than one agency. Some of the examples of overlapping schemes are as follows:

- Several areas addressed inadequately: There are a host of areas which are not addressed by these schemes such as financing of freezer cabinets for retail outlets, financing of bulkcoolers for milk, financing of alternate markets/mandis, financing of vending machines for tea/coffee/beverages etc.
- Creation of excess capacity: There are instances where assistance has been provided to set up cold storages in locations which have excess storage capacity. This has lead to an unviable situation for all cold storage operators in the region. It is essential for the nodal agencies to achieve development of the food processing sector in a sustainable manner.
- Inadequate financial assistance^ most instances, there is a cap on total financial assistance provided at INR 5-5.5 mn. This leads to fragmentation in capacity creation, and often, a situation of overcapacity. At the same time, it does not allow players to scale up their operations.
- Inadequate Monitoring of progress of schemes: It is important to track progress of the projects funded by the Government. Periodic appraisal of the progress will ensure that funds are appropriately utilized. This will also enable assessment of success and lacunae of these schemes and record the multiplier effect of the financial support provided.
- Most of the schemes are back-ended which leads to a funds crunch during implementation.
- Institutions providing subsidies do not undertake their own assessment and rely on the appraisal report of the bank which provides loans to the project.
- There are significant time lags from the date of application for financial assistance, to release of funds, and may affect the project schedule, and lead to cost overruns
- These schemes do not address working capital requirements, which due to the nature of the industry are significant and often higher in terms of quantum than capital investment requirements
- Applicants claim that they are often unaware that their applications have been rejected on account of lack of communication from the funding institution. Further, applicants are unaware of the reasons for rejection.

KEY ISSUES IN FINANCING: FOOD PROCESSORS' PERSPECTIVE

With raw material availability often being seasonal, and/or concentration in demand, inventory holdings are high and thus the working capital finance issued through normal Maximum Permissible Bank Finance method does not address the funding requirements adequately. The rate of interest for working capital assistance is high. Most food processing companies are not able to access adequate working capital at reasonable rates. This affects the raw material procurement and capacity utilization through out the year. Unavailability of large working capital facilities, or facilities without a high seasonal drawal, restricts players in purchasing agri produce in large volumes, when prices are favourable.

KEY ISSUES IN FINANCING FOOD PROCESSORS: BANKER'S PERSPECTIVE

- Lack of reliable information on demand-supply, price trends, raw material supplies etc is a significant constraint for banks, impacting credit assessment and monitoring
- High operational/transaction cost in servicing small companies
- Lack of backward linkages for assured access to raw material
- Lack of forward linkages of stand-alone food processing units for marketing/distribution

Owing to high cost of market development, the profitability of food processing companies is under pressure in the initial years of their operation. This impacts the risk rating and leads to high cost of borrowing. Even temporary adversities in market conditions can lead to defaults and eventually the loan can become a non performing asset for banks.

FDI IN FOOD PROCESSING INDUSTRIES

FDI in the food processing sector is low, constituting about 4% of total FDI inflow during the period 1991 to 2004. Most food processing sectors have been placed under a liberal, transparent and investor friendly FDI policy allowing 100% FDI through the automatic route for most food processing sectors. However, some of the sub-sectors which are barred from FDI include agriculture and plantation excluding tea plantations and retail trade in any form (except cash & carry formats). In alcoholic beverages setting up of new manufacturing capacities is not permitted and only existing players can expand capacities, which hinders FDI flow to the sector.

ISSUES IN FDI FOR FOOD PROCESSING

1 Supply-chain issues: Factors such as lack of farmer-processor linkages, poor quality of infrastructure, fragmented retail distribution and regulatory hurdles, which are discussed in earlier sections of this report which have affected inflows of FDI in food processing.

2 Procedural bottlenecks: The requirement of multiple approvals, together with delays in obtaining approvals leads to cost overruns, and impacts investor confidence. A project can be categorised into three stages-approval, clearance and implementation. Industry sources state that the stage of obtaining clearances is a bottleneck, as there is duplication of procedures of the Central and the State Governments.

3. Quality perception: While this perception is fast changing for other sectors such as pharmaceuticals, foreign investors associate India with low/average quality in agricultural produce.

In contrast to India, China has seen significant FDI inflows, including in food processing. Sectors, such as soft drinks, infant formula milk powder, instant noodle, and snacks. Key international players with significant investments in China include: Danone, the Coca-Cola Co., PepsiCo, Cadbury, Hormel, Nestle, Unilever, Wyeth, Mead Johnson, Want Want and Uni-President. Some of the measures which have augmented FDI inflows in food processing are:

- Single window clearing system for Agri/food projects
- Development of agri processing zones near coastal areas
- State investment in development of logistics systems
- Land reforms-Removal of collective land ownership system, extending land lease to 30 years for individual households. Investment in high yielding varieties, across crops. Upgrading cultivation practices in states to cater to specific

world markets *e.g.* Shandong province shifted from grains to vegetables to cater to the Japanese market

Recommendations to Increase FDI in Food Processing

In addition to implementing solution themes to enhance supply chain efficiencies and remove regulatory bottlenecks, specific measures to increase FDI in food processing include:

- Single window clearance for FDI in Food processing can assist in increasing inflow into the sector. Association of a Central government officer with the investor, who facilitates all clearances, both Central and State, can be an effective solution
- It is imperative for both the Central and the State Government to launch a high intensity campaign for investments in the food processing sector in India. With the reduction of agricultural subsidies across several developed markets, international players will seek to invest in high growth markets. Market development efforts can therefore boost the confidence of potential entrants. A sector specific approach for marketing to international investors is required. Also, examples of successful investments by MNCs can be showcased to international companies. The MFPI can coordinate these efforts together with respective State Governments.

PICKLING

Pickling, also known as brining or corning is the process of preserving food by anaerobic fermentation in brine to produce lactic acid, or marinating and storing it in an acid solution, usually vinegar. The resulting food is called a pickle. This procedure gives the food a salty or sour taste. In South Asia, edible oils are used as the pickling medium with vinegar. Another distinguishing characteristic is a pH less than 4.6, which is sufficient to kill most bacteria. Pickling can preserve perishable foods for months. Antimicrobial herbs and spices, such as mustard seed, garlic, cinnamon or cloves, are often added. If the food contains sufficient moisture, a pickling brine may be produced simply by adding dry salt. For example, sauerkraut and Korean kimchi are produced by salting the vegetables to draw out excess water. Natural fermentation at room temperature, by lactic acid bacteria, produces the required acidity. Other pickles are made by placing vegetables in vinegar. Unlike the canning process, pickling does not require that the food be completely sterile before it is sealed. The acidity or salinity of the solution, the temperature of fermentation, and the exclusion of oxygen determine which microorganisms dominate, and determine the flavour of the end product. When both salt concentration and temperature are low, *Leuconostoc mesenteroides* dominates, producing a mix of acids, alcohol, and aroma compounds. At higher temperatures *Lactobacillus plantarum* dominates, which produces primarily lactic acid. Many pickles start with Leuconostoc, and change to Lactobacillus with higher acidity.

PICKLE ETYMOLOGY AND HUMOUR

The term pickle is derived from the Dutch word pekel, meaning brine. In the US and Canada, the word pickle alone almost always refers to a pickled cucumber, except when it is used figuratively. Pickle is considered an inherently funny word. In Neil Simon's play *The Sunshine Boys*, a character explains, "Words with a k in it are funny. Alka-Seltzer is funny. Chicken is funny. Pickle is funny...." For example, the US Central Intelligence Agency is nicknamed "the pickle factory", When plans for the Chicago Picasso sculpture were first unveiled, to public derision, a waggish publicist erected a giant dill pickle statue on the proposed site.

THE PICKLING PROCESS

The jar and lid are first boiled in order to sterilize them. The fruits or vegetables to be pickled are then added to the jar along with brine and/or vinegar and are then allowed to ferment until the desired taste is obtained.

METHOD OF VACUUM PACKING

Vacuum packing is a method of storing food and presenting it for sale. Appropriate types of food are stored in an airless environment, usually in an air-tight pack or bottle to prevent the growth of microorganisms. The vacuum environment removes atmospheric oxygen, protecting the food from spoiling by limiting the growth of aerobic bacteria or fungi, and preventing the evapouration of volatile components. Vacuum packing is commonly used for long-term storage of dry foods such as cereals, nuts, cured meats, cheese, smoked fish, coffee, and potato chips.

It is also for storage of fresh foods such as vegetables, meats, and liquids such as soups in a shorter term because vacuum condition cannot stop bacteria from getting water which can promote their growth. Vacuum packaging food can extend its life by up to 3-5 times. Vacuum packing is also used to reduce greatly the bulk of non-food items. For example, clothing and bedding can be stored in bags evacuated with a domestic vacuum cleaner or a dedicated vacuum sealer.

This technique is sometimes used to compact household waste, for example where a charge is made for each full bag collected. Vacuum packing can be used to reduce bulk of inflatable items as well. Vacuum packaging products using plastic bags, canisters, bottles, or mason jars are available for home use. Vacuum packaging delicate food items can be done by using an inert gas kit, typically available on chamber vacuum sealers. After air has been removed, an inert gas is added to maintain the preservation of packaged food while preventing damage. An example of inert gas for packaging delicate foods is potato chips. External vacuum sealers involve a bag being attached to the vacuum-sealing machine externally. The machine will remove the air and seal the bag, which is all done outside the machine. Chamber sealers require the entire product to be placed within the machine. Like external sealers, a plastic bag is typically used for packaging. Once the product is placed in the machine, the lid is closed and air is removed. Once the air is removed, the bag is sealed and the atmosphere within the chamber is returned back to normal. The lid is then opened and the product removed. Chamber sealers are typically used for higher-volume packaging. Manufacturers of chamber type vacuum packing machines include: Cryovac, Multivac, Sammic, VC999, Sevana and several others. When foods are frozen without preparation, freezer burn can occur. It happens when the surface of the food is dehydrated, and this leads to a dried and leathery appearance.

Freezer burn also ruins the flavour and texture of foods. Vacuum packing prevents freezer burn by preventing the food from exposure to the cold, dry air.Vacuum packaging also allows for a special cooking method, Sous-vide. Sous-vide, meaning "under vacuum" in French, involves poaching food that is vacuum sealed in a plastic bag. Due to an oxygen-poor environment, anaerobic microorganism can proliferate, so vacuum packing is often used in combination with other treatment.

DRYING

Drying is a method of food preservation that works by removing water from the food, which inhibits the growth of microorganisms and hinders quality decay. Drying food using sun and wind to prevent spoilage has been practised since ancient times. Water is usually removed by evapouration but, in the case of freeze-drying, food is first frozen and then the water is removed by sublimation. Bacteria yeasts and moulds need the water in the food to grow. Drying effectively prevents them from surviving in the food.

FOOD TYPES

Many different foods are prepared by dehydration. Good examples are meat such as prosciutto, bresaola, and beef jerky. Dried and salted reindeer meat is a traditional Sami food. First, the meat is soaked/ pickled in saltwater for a couple of days to guarantee the conservation of the meat. Then the meat is dried in the sun in spring when the air temperature is below zero.

The dried meat can be further processed to make soup. Fruits change character completely when dried: the plum becomes a prune, the grape a raisin; figs and dates are also transformed in new, different products, that can be eaten as they are or else after rehydration. Home drying of vegetables, fruit and even meat may be carried out by a do-it-yourself practice, employing electrical dehydrators. If the user does not like to use additives as potassium metabisulphite, or BHA, BHT for meats, dried products may be hermetically shelf stored if it is to be consumed soon, or else in the refrigerator or even freezer if a long storage is to be expected.

Freeze dried vegetables are often found in backpackers food, hunters, military, etc. The exception to this rule are bulbs, such as garlic and onion, which are often dried. Also chilis are frequently dried. Edible and psilocybin mushrooms, as well as other fungi, are also sometimes dried for preservation purposes, to affect the potency of chemical components, or so they can be used as seasonings. For centuries, much of the European diet depended on dried cod, known as salt cod or bacalhau or stockfish. It formed the main protein source for the slaves on the West Indian plantations, and was a major economic force within the triangular trade. Dried shark meat, known as Hákarl, is a delicacy in Iceland.

GRAIN DRYING

Hundreds of millions of tonnes of wheat, corn, soybean, rice and other grains as sorghum, sunflower seeds, rapeseed/canola, barley, oats, etc., are dried in grain dryers. In the main agricultural countries, drying comprises the reduction of moisture from about 17-30 per cent w/w to values between 8 and 15 per cent w/w, depending on the grain. The final moisture content for drying must be adequate for storage. The more oil the grain has, the lower its storage moisture content will be. Cereals are often dried to 14 per cent w/w, while oilseeds, to 12.5 per cent (soybeans), 8 per cent (sunflower) and 9 per cent (peanuts). Drying is carried out as a requisite for safe storage, in order to inhibit microbial growth. However, low temperatures in storage are also highly recommended to avoid degradative reactions and, especially, the growth of insects and mites.

A good maximum storage temperature is about 18°C. The largest dryers are normally used "Off-farm", in elevators, and are of the continuous type: Mixed-flow dryers are preferred in Europe, while Cross-flow dryers in the USA. In Argentina, both types are usually found. Continuous flow dryers may produce up to 100 metric tonnes of dried grain per hour.

The depth of grain the air must traverse in continuous dryers range from some 0.15 m in Mixed flow dryers to some 0.30 m in Cross-Flow. Batch dryers are mainly used "On-Farm", particularly in the USA and Europe. They normally consist of a bin, with heated air flowing horizontally from an internal cylinder through an inner perforated metal sheet, then through a annular grain bed, some 0.50 m thick in radial direction, and finally across the outer perforated metal sheet, before being discharged to the atmosphere.

The usual drying times range from 1 h to 4 h depending on how much water must be removed, type of grain, air temperature and the grain depth. In the USA, continuous counterflow dryers may be found on-farm, adapting a bin to slowly drying grain fed at the top and removed at the bottom of the bin by a sweeping auger. Grain drying is an active area of manufacturing and research. Now it is possible to simulate the performance of a dryer with computer programmes based on equations that represent the phenomena involved in drying: physics, physical chemistry, thermodynamics and heat and mass transfer.

Most recently the evolution of quality indices is beginning to be predicted with some confidence, in order to add an essential performance parametre with which to establish a compromise of reasonably fast drying rate, limited energy consumption, and satisfactory grain quality. A typical quality parametre in wheat drying is the breadmaking quality and germination percentage whose reductions in drying are somewhat related.

FINANCING - FOOD PROCESSING SECTOR

The food processing sector comprises a large number of small and medium sized companies, a significant proportion of which have stand alone operations, with no control over the raw material base and reliant on other organizations to undertake marketing/further processing of their products. Banks and financial Institutions adopt the same risk models relevant to the manufacturing sector, for assessing food processing companies. Interest charges for food processing companies are high, on account of the high risk perception associated with the nature of their operations.

OTHER FOOD MANUFACTURING/ PROCESSING OPERATIONS

The Food Code specifies under that the food establishment operator must obtain a variance from the regulatory authority for all food manufacturing/processing operations based on the prior approval of a HACCP plan.

The purpose of this is to provide processing criteria for different types of food manufacturing/processing operations for use by those preparing and reviewing HACCP plans and proposals. Criteria for additional processes will be provided as they are developed, reviewed, and accepted.

REDUCED OXYGEN PACKAGING (ROP)

ROP which provides an environment that contains little or no oxygen, offers unique advantages and opportunities for the food industry but also raises many microbiological concerns. Products packaged using ROP may be produced safely if proper controls are in effect. Producing and distributing these products with a HACCP approach offer an effective, rational, and systematic method for the assurance of food safety. The purpose of this Annex is to provide guidelines for effective food safety controls for retail food establishments covering the receipt, processing, packaging, holding, displaying, and labeling of food in reduced oxygen packages.

DEFINITIONS

The term ROP is defined as any packaging procedure that results in a reduced oxygen level in a sealed package.

The term is often used because it is an inclusive term and can include other packaging options such as:

- Cook-chill is a process that uses a plastic bag filled with hot cooked food from which air has been expelled and which is closed with a plastic or metal crimp.
- Controlled Atmosphere Packaging is an active system which continuously maintains the desired atmosphere within a package throughout the shelf-life of a product by the use of agents to bind or scavenge oxygen or a sachet containing compounds to emit a gas. Controlled Atmosphere Packaging is defined as packaging of a product in a modified atmosphere followed by maintaining subsequent control of that atmosphere.
- Modified Atmosphere Packaging is a process that employs a gas flushing and sealing process or reduction of oxygen through respiration of vegetables or microbial action. Modified Atmosphere Packaging is defined as packaging of a product in an atmosphere which has had a one-time modification of gaseous composition so that it is different from that of air, which normally contains 78.08 per cent nitrogen, 20.96 per cent oxygen, 0.03 per cent carbon dioxide.

- Sous Vide is a specialized process of ROP for partially cooked ingredients alone or combined with raw foods that require refrigeration or frozen storage until the package is thoroughly heated immediately before service. The sous vide process is a pasteurization step that reduces bacterial load but is not sufficient to make the food shelf-stable.

  The process involves the following steps:
  - Preparation of the raw materials;
  - Packaging of the product, application of vacuum, and sealing of the package;
  - Pasteurization of the product for a specified and monitored time/temperature;
  - Rapid and monitored cooling of the product at or below 3°C or frozen; and
  - Reheating of the packages to a specified temperature before opening and service.
- Vacuum Packaging reduces the amount of air from a package and hermetically seals the package so that a near-perfect vacuum remains inside. A common variation of the process is Vacuum Skin Packaging. A highly flexible plastic barrier is used by this technology that allows the package to mold itself to the contours of the food being packaged.

TRENDS IN FOOD PACKAGING

Numerous reports of industry associations agree that use of smart indicators will increase. There are a number of different indicators with different benefits for food producers, consumers and retailers.

TEMPERATURE RECORDERS

Temperature recorders are used to monitor products shipped in a cold chain and to help validate the cold chain. Digital temperature data loggers measure and record the temperature history of food shipments.

They sometimes have temperatures displayed on the indicator or have other output (lights, etc.): The data from a shipment can be downloaded (cable, RFID, etc.) to a computer for further analysis.

These help identify if there has been temperature abuse of products and can help determine the remaining shelf life. They can also help determine the time of temperature extremes during shipment so corrective measures can be taken.

Time-Temperature Indicators

Time-Temperature Indicators integrate the time and temperature experienced by the indicator and adjacent foods. Some use chemical reactions that result in a colour change while others use the migration of a dye through a filter media. To the degree that these physical changes in the indicator match the degradation rate of the food, the indicator can help indicate probable food degradation.

RFID

Radio Frequency Identification is applied to food packages for supply chain control and have shown a significant benefit in allowing food producers and retailers create full real time visibility of their supply chain.

FUNCTIONS OF FOOD PACKAGING

Packaging has several objectives:

- *Physical Protection*: The food enclosed in the package may require protection from, among other things, shock, vibration, compression, temperature, etc.
- *Barrier Protection*: A barrier from oxygen, water vapour, dust, etc., is often required. Permeation is a critical factor in design. Some packages contain desiccants or oxygen absorbers to help extend shelf life. Modified atmospheres or controlled atmospheres are also maintained in some food packages. Keeping the contents clean, fresh, and safe for the intended shelf life is a primary function.
- *Containment or Agglomeration*: Small items are typically grouped together in one package for reasons of efficiency. Powders, and granular materials need containment.
- *Information Transmission*: Packages and labels communicate how to use, transport, recycle, or dispose of the package or product. Some types of information are required by governments.
- *Marketing*: The packaging and labels can be used by marketers to encourage potential buyers to purchase the

product. Package design has been an important and constantly evolving phenomenon for several decades. Marketing communications and graphic design are applied to the surface of the package and (in many cases) the point of sale display.

- *Security*: Packaging can play an important role in reducing the security risks of shipment. Packages can be made with improved tamper resistance to deter tampering and also can have tamper-evident features to help indicate tampering. Packages can be engineered to help reduce the risks of package pilferage: Some package constructions are more resistant to pilferage and some have pilfer indicating seals. Packages may include authentication seals to help indicate that the package and contents are not counterfeit. Packages also can include anti-theft devices, such as dye-packs, RFID tags, or electronic article surveillance tags, that can be activated or detected by devices at exit points and require specialised tools to deactivate. Using packaging in this way is a means of retail loss prevention.
- *Convenience*: Packages can have features which add convenience in distribution, handling, stacking, display, sale, opening, reclosing, use, and reuse.
- *Portion Control*: Single serving packaging has a precise amount of contents to control usage. Bulk commodities (such as salt) can be divided into packages that are a more suitable size for individual households. It also aids the control of inventory: selling sealed one-litre-bottles of milk, rather than having people bring their own bottles to fill themselves.

FOOD PACKAGING TYPES

The above materials are fashioned into different types of food packages and containers such as:

Packaging Type	Type of Container	Food Examples
Aseptic processings	Primary	Liquid whole eggs

Plastic trays	Primary	Portion of fish
Bags	Primary	Potato chips
Boxes	Secondary	Box of Coca-Cola
Cans	Primary	Can of Campbell's tomato soup
Cartons	Primary	Carton of eggs
Flexible packaging	Primary	Bagged salad
Pallets	Tertiary	A series of boxes on a single pallet used to transport from the manufacturing plant to a distribution centre
Wrappers	Tertiary	Used to wrap the boxes on the pallet for transport

Primary packaging is the main package that holds the food that is being processed. Secondary packaging combines the primary packages into one box being made. Tertiary packaging combines all of the secondary packages into one pallet.

There are also special containers that combine different technologies for maximum durability:

- Bags-in-Boxes (used for soft drink syrup, other liquid products, and meat products).
- Wine box (used for wine).

PACKAGING MACHINES

A choice of packaging machinery includes technical capabilities, labour requirements, worker safety, maintaina-bility, serviceability, reliability, ability to integrate into the packaging line, capital cost, floorspace, flexibility (change-over, materials, etc.), energy usage, quality of outgoing packages, qualifications (for food, phamaceuticals, etc.), throughput, efficiency, productivity, ergonomics, etc.

Packaging machines may be of the following general types:

- Blister, Skin and Vacuum Packaging Machines.
- Capping, Over-Capping, Lidding, Closing, Seaming and Sealing Machines.
- Cartoning Machines.
- Case and Tray Forming, Packing, Unpacking, Closing and Sealing Machines.

- Check Weighing Machines.
- Cleaning, Sterilizing, Cooling and Drying Machines.
- Conveying, Accumulating and Related Machines.
- Feeding, Orienting, Placing and Related Machines.
- *Filling Machines*: Handling liquid and powdered products.
- Package Filling and Closing Machines.
- Form, Fill and Seal Machines.
- Inspecting, Detecting and Checkweighing Machines.
- Palletizing, Depalletizing, Pallet Unitizing and Related Machines.
- *Product Identification*: Labelling, marking, etc.
- Wrapping Machines.
- Converting Machines.
- Other speciality machinery.

REDUCING FOOD PACKAGING

Reduced packaging and sustainable packaging are becoming more frequent. The motivations can be government regulations, consumer pressure, retailer pressure, and cost control. (Reduced packaging often saves packaging costs.) In the UK, a Local Government Association survey produced by the British Market Research Bureau, compared a range of outlets to buy 29 common food items, found that small local retailers and market traders "produced less packaging and more that could be recycled than the larger supermarkets".

Index